Climate in Peril

A popular guide to the latest IPCC reports

Disclaimer: This guide tries to remain faithful to the sense of the work of the IPCC and its Climate Change 2007 Synthesis Report of the Fourth Assessment, while simplifying the language and structure. The full responsibility for the accuracy of the content rests with the editors of this guide. Readers may remember the IPCC's helpful warning that "while the Synthesis Report is a largely self-contained document, it needs to be viewed in the context of the other volumes of *Climate Change 2007* and it is recommended that for further details the contributions of the three Working Groups be consulted, published in the volumes "Climate Change 2007 – The Physical Science Basis"; "Impacts, Adaptation and Vulnerability" and "Mitigation of Climate Change", as well as "Climate Change 2007 Synthesis Report".

6 Robust findings and key uncertainties
8 Present changes, causes and observed impacts
24 Projected climate change and its impacts
41 Adaptation and mitigation
54 Sustainable development, environmental protection and climate change
55 Key vulnerabilities, impacts and risks – longterm perspectives

Text
Alex Kirby

GRID-Arendal/GENFSEC editorial team
Christina Stuhlberger
Claudia Heberlein

Senior expert and initiator of the publication
Svein Tveitdal, Klima 2020

Carto-graphics
Viktor Novikov, GRID-Arendal/GENFSEC
Matthias Beilstein

English editing of graphics
Harry Forster, Interrelate Grenoble

Layout
GRID-Arendal

Foreword

In 2007 the Intergovernmental Panel on Climate Change (IPCC) shared the Nobel Peace Prize with former US Vice President Al Gore for their work to provide policy makers and the general public around the world with the best possible science base for understanding and combating the increasing threat from climate change. But as the messages from the scientists are becoming increasingly explicit, the gap between the need for action they project and the climate policy the world leaders put in place is steadily increasing.

One illustration is the trend in emissions of greenhouse gases. According to the IPCC global emissions would need to peak between 2000 and 2015 in order to limit the global temperature increase to between 2 and 2.4°C compared to pre-industrial times. In 2007, when ideally the emissions should have peaked, the world instead experienced a new record in annual emission increase. For each day we fail to twist development towards a low-carbon society, the damage to the world's ecosystems become more severe, and the costs of mitigation and adaptation increases.

The main purpose of this short guide is to help bridging the gap between science and policy and to increase public awareness about the urgency of action to combat climate change and its impacts. This booklet is intended for those who do not have the time – and may not have the scientific expertise – to read the entire Synthesis Report from the IPCC.

Special thanks to Svein Tveitdal of Klima 2020 and former director of Grid-Arendal, for the initiative for this booklet and his valuable contribution to the content. We would also like to express our gratitude to the Swedish Environment Protection Agency and Earth-Print Ltd for additional financial support.

Arendal and Oslo, February 10, 2009

Peter Prokosch,
Director,
GRID-Arendal

Ellen Hambro,
Director General,
Norwegian Pollution
Control Authority (SFT)

How to Use This Guide

This guide, while it aims to present the substance and the sense of the IPCC's original Synthesis Report, is designed to be read as a narrative. So it tells the story in a simplified language while taking the liberty of shortening or enhancing specific parts where it appears useful and illustrating the text with additional graphics. You will always find the source of the data mentioned if it differs from the IPCC's own. The guide covers the six original topic headings as in the Summary for Policymakers but the order in which they are presented here differs from the IPCC publication. It starts by spelling out what the IPCC knows and what it considers as key questions.

Although the guide is intended for lay readers, not climate scientists, inevitably it uses some scientific terms. Readers will find a fuller explanation of some of them in the short Glossary at the end of the guide: they appear in the text in *italics*. In their assessment reports, the IPCC uses commonly used terms with a very specific meaning. In order to simplify the language, this guide abandon these specialized terms. *The IPCC also uses several terms which are likely to be self-explanatory: they include high agreement/medium agreement and high evidence/medium evidence*. The term *"agreement"* refers to agreement found within the scientific literature.

Introduction

In 2007 the Intergovernmental Panel on Climate Change published its Fourth Assessment Report (following earlier assessment reports in 1990, 1995 and 2001). The report – AR4 for short – consists of four volumes, published under the title *Climate Change 2007*. One volume was devoted to each of the IPCC's three Working Groups:

Working Group I (WG I) assesses the physical scientific aspects of the climate system and climate change;
Working Group II (WG II) assesses the vulnerability of socio-economic and natural systems to climate change, negative and positive consequences, and options for adapting to it; and
Working Group III (WG III) assesses options for mitigating climate change through limiting or preventing greenhouse gas emissions and enhancing activities that are working to remove them from the atmosphere.

The fourth volume that completed AR4 is the Synthesis Report. It summarizes the findings of the other three volumes and specifically addresses issues of concern to policymakers, and draws as well on other IPCC reports. Its range of policy-relevant questions is structured around six topic headings:

1. Observed changes in climate, and the effects of past changes
2. Natural and human causes of climate change and their relation to observed changes
3. Projected future climate change and its impacts
4. Options to adapt to climate change and to mitigate it; what responses are possible by 2030
5. The long-term perspective; how fast and how deep greenhouse gas cuts will need to be to limit global temperature increases to a given level; why climate concerns are intensifying
6. Robust findings and key uncertainties.

The IPCC is a scientific intergovernmental body set up by the World Meteorological Organization (WMO) and by the United Nations Environment Programme (UNEP) in 1988. It was established to provide decision-makers and others interested in climate change with an objective source of information. The IPCC does not conduct any research. Its role is to assess on a comprehensive, objective and transparent basis the latest scientific, technical and socio-economic literature relevant to the understanding of the risk of human-induced climate change, its observed and projected impacts and options for adaptation and mitigation. IPCC reports should be neutral with respect to policy, although they need to deal objectively with policy relevant scientific, technical and socio-economic factors. They should aim to reflect a range of views, expertise and wide geographical coverage. The IPCC continues to be a major source of information for the negotiations under the UNFCCC (United Nations' Framework Convention on Climate Change).

How the IPCC is organized

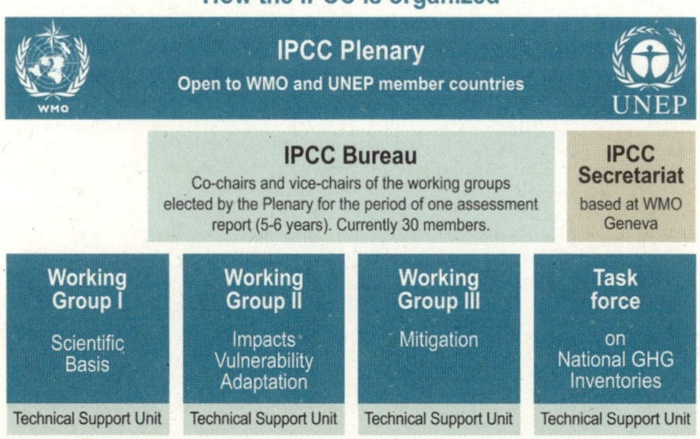

Robust findings and key uncertainties

Robust findings

Observed changes in climate, their effects and their causes

- Warming is un ambiguous, as demonstrated by observations such as:
  - rises in global average air and sea temperatures and average sea levels,
  - widespread melting of snow and ice;
- Observed changes in many biological and physical systems are consistent with warming:
  - many natural systems on all continents and oceans are affected;
- 70% growth of greenhouse gas emissions in terms of the global warming potential between 1970 and 2004;
- Concentrations of methane (CH_4), carbon dioxide (CO_2) and nitrous oxide (N_2O) are now far higher than their natural range over many thousands of years before industrialization (1750);
- Most of the warming over the last 50 years is "very likely" to have been caused by anthropogenic increases in greenhouse gases.

Causes and projections of future climate changes and their impacts

- Global GHG emissions will continue to grow over decades unless there are new policies to reduce climate change and to promote sustainable development.
- Warming of about 0.2°C a decade is projected for the next two decades (several IPCC scenarios).
- Changes this century would "very likely" be larger than in the 20th.
- Greater warming over land than sea, and more in the high latitudes of the northern hemisphere.
- The more the planet warms, the less CO_2 it can absorb naturally.
- Warming and rising sea levels would continue for centuries, even if GHG emissions were reduced and concentrations stabilized, due to feedbacks and the time-lag between cause and effect.
- If GHG levels in the atmosphere double compared with pre-industrial levels, it is "very unlikely" that average global temperatures will increase less than 1.5°C compared with that period.

Responses to climate change

- Some planned adaptation to climate change is occurring, but much more is needed to reduce vulnerability.
- Long term unmitigated climate change will "likely" exceed the capacity of people and the natural world to adapt.
- Many technologies to mitigate climate change are already available or likely to be so by 2030. But incentives and research are needed to improve performance and cut cost.
- The economic mitigation potential, at costs from below zero to US$100 per tonne of CO_2 eq, is enough to offset the projected growth of global emissions or to cut them to below their current levels by 2030.
- Prompt mitigation can buy time to stabilize emissions and to reduce, delay or avoid impacts.
  Sustainable development and appropriate policy-making in sectors not apparently linked to climate help to stabilize emissions.
  Delayed emissions reductions rincrease the risk of more severe climate change impacts.

Key uncertainties

Observed changes in climate, their effects and their causes

- Limited climate data coverage in some regions.
- Analysing and monitoring changes in extreme events, for example droughts, tropical cyclones, extreme temperatures and intense precipitation (rain, sleet and snow), is harder than identifying climatic averages, because longer and more detailed records are needed.
- Difficult to determine the effects of climate change on people and some natural systems, because they may adapt to the changes and because other unconnected causes may be exerting an influence.
- Hard to be sure, at scales smaller than an entire continent, whether natural or human causes are influencing temperatures because (for example) pollution and changes in land use may be responsible.
- There is still uncertainty about the scale of CO_2 emissions due to changes in landuse, and the scale of methane emissions from individual sources.

Causes and projections of future climate changes and their impacts

- It is uncertain how much warming will result in the long term from any particular level of GHG concentrations, and therefore,
- it is uncertain what level - and pace - of emissions cuts will be needed to ensure a specific level of GHG concentrations.
- Estimates vary widely for the impacts of aerosols and the strength of feedbacks, particularly clouds, heat absorption by the oceans, and the carbon cycle.
- Possible future changes in the Greenland and Antarctic ice sheets are a major source of uncertainty about rising sea levels
- Projections of climate change impacts beyond about 2050 are heavily dependent on scenarios and models.

Responses to climate change

- Limited understanding of how development planners factor climate into their decisions.
- Effective adaptation steps are highly specific to different political, financial and geographical circumstances, making it hard to appreciate their limitations and costs.
- Estimating mitigation costs and potentials depends on assumptions about future socio-economic growth, technological change and consumption patterns.
- Not enough is known about how policies unrelated to climate will affect emissions.

Present changes, causes and observed impacts

Observed changes in climate and their effects

Warming of the climate system is beyond argument, as shown by observations of increases in global average air and ocean temperatures, the widespread melting of snow and ice, and rising global average sea levels. In the following paragraphs, some of the most striking changes that are already taking place are described and illustrated.

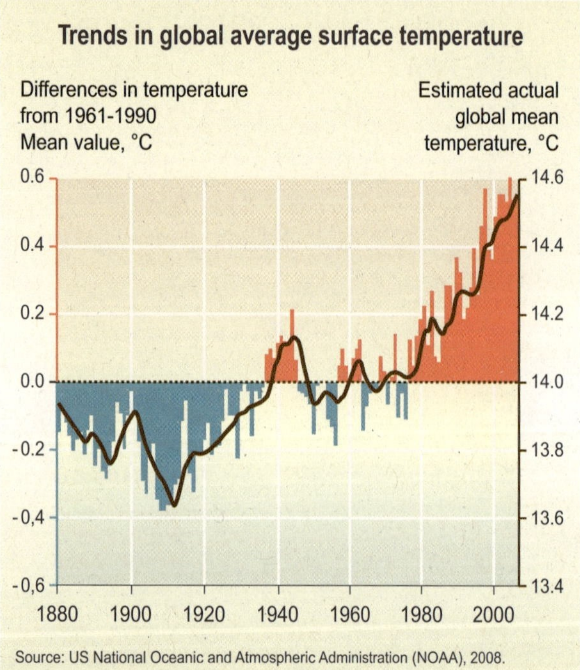

Source: US National Oceanic and Atmospheric Administration (NOAA), 2008.

Temperature rise

Of the last 12 years (1995–2006), 11 are among the 12 warmest since records began in 1850. The warming trend over the previous century was reported as 0.6°C in the IPCC's Third Assessment Report (TAR) published six years earlier in 2001: it is now 0.74°C. The temperature increase is widespread across the world but is most marked in the northern Polar Regions. Warming of the climate system has been detected on the Earth's surface and up in the atmosphere, as well as in the upper few hundred metres of the oceans. The land has warmed faster than the seas.

Average Northern Hemisphere temperatures after 1950 have been higher than during any other 50-year

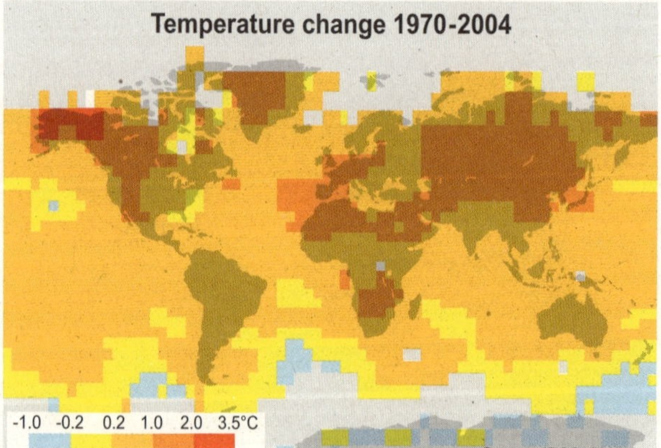

Source: IPCC, 2007.

period in the last 500 years. These rising temperatures unavoidably have an influence on very diverse natural phenomena that we so far have taken for granted.

Evidence of a warming world includes shorter freezing seasons of lake and river ice, decreases in the extent of permafrost, and rising soil temperatures. But the main changes observed scientifically, but also noticed more and more by the people all over the world are summarized in the following paragraphs.

Sea level rise

Sea levels across the globe have risen in a way consistent with the warming – since 1961 at an average of 1.8 millimetres per year, and since 1993 at 3.1 millimetres per year. Scientists are not sure whether to attribute this last decade's higher rise to a variation from one decade to another, or whether it marks a longer-term trend. The total global rise in the 20th century amounted to 17 centimetres. The expansion

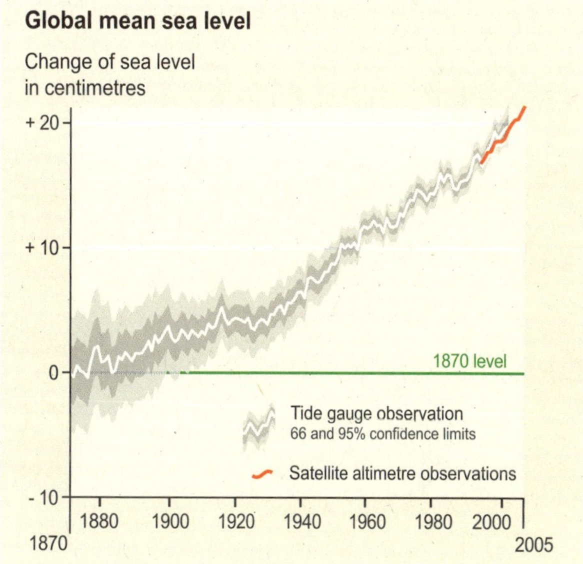

Source: Hugo Ahlenius, GRID-Arendal 2008, updated from Church and White 2006.

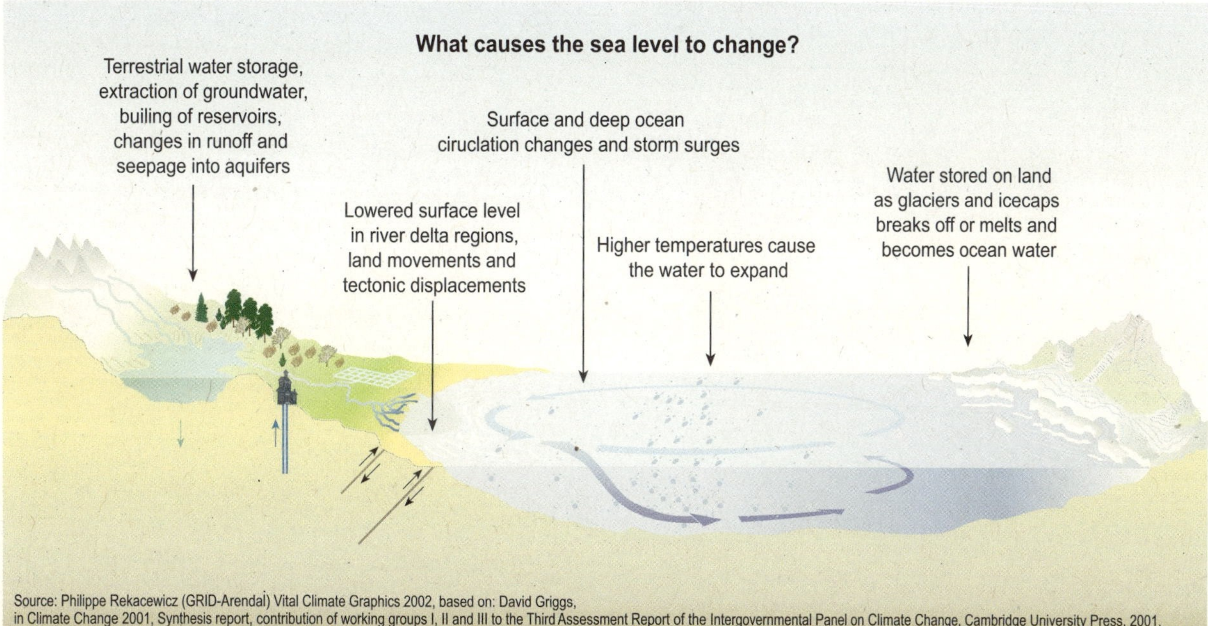

Source: Philippe Rekacewicz (GRID-Arendal) Vital Climate Graphics 2002, based on: David Griggs, in Climate Change 2001, Synthesis report, contribution of working groups I, II and III to the Third Assessment Report of the Intergovernmental Panel on Climate Change, Cambridge University Press, 2001.

of water as it warms, and the melting of glaciers, ice caps and the polar ice sheets are all contributing to the rise.

Melting snow and ice

Decreases in snow and ice extent are also consistent with warming. Satellite data recorded since 1978 show the annual average Arctic sea ice extent has shrunk by 2.7 per cent each decade, with larger decreases in summer. Mountain glaciers and average snow cover have declined in both hemispheres.

Extreme weather events

From 1900 to 2005 precipitation (rain, sleet and snow) increased significantly in parts of the Americas, northern Europe and northern and central Asia, but declined in the Sahel, the Mediterranean, southern Africa and parts of southern Asia. The IPCC concludes that it is likely that the global area affected by drought has increased since the 1970s.

Cold days, nights and frosts have become less frequent over most land areas in the past 50 years, and hot days and nights more frequent. The IPCC considers that it is likely that heat waves have become commoner over most land areas, that heavy precipitation events (thunderstorms, for instance) have increased over most areas, and that since 1975 extreme high sea levels have increased worldwide – in addition to the rise in average sea levels.

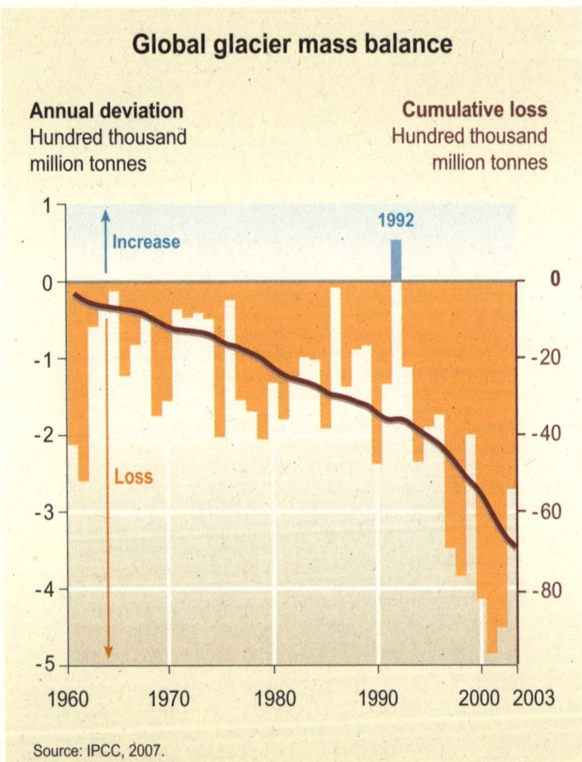

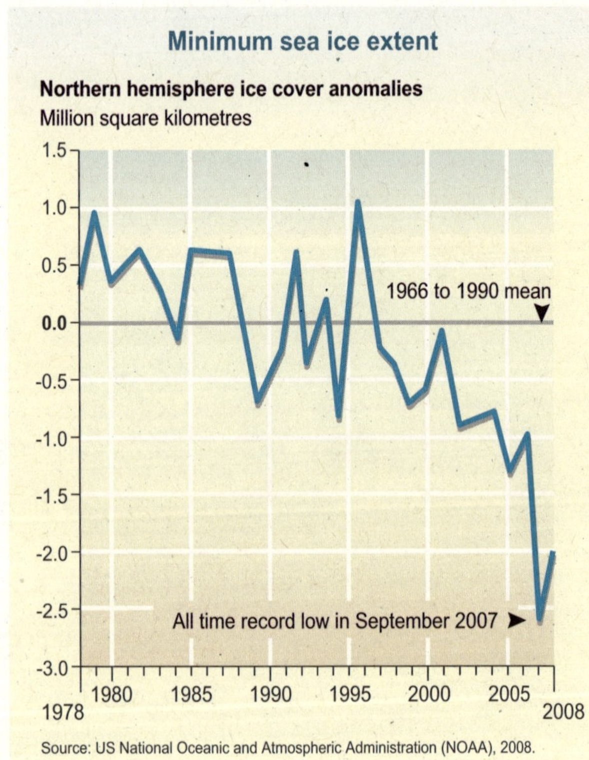

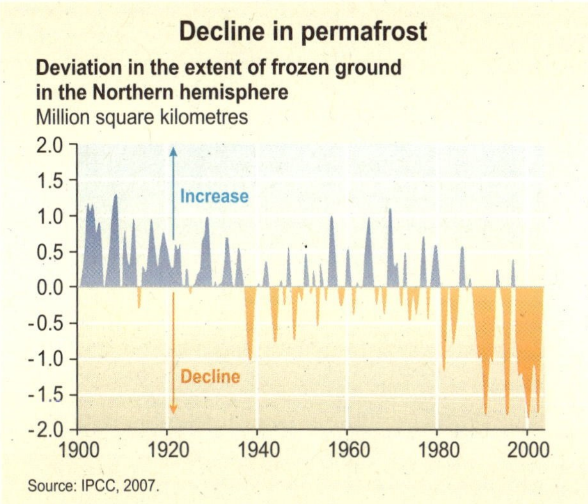
Decline in permafrost
Deviation in the extent of frozen ground in the Northern hemisphere
Million square kilometres
Increase
Decline
2.0
1.5
1.0
0.5
0.0
-0.5
-1.0
-1.5
-2.0
1900
1920
1940
1960
1980
2000
Source: IPCC, 2007.

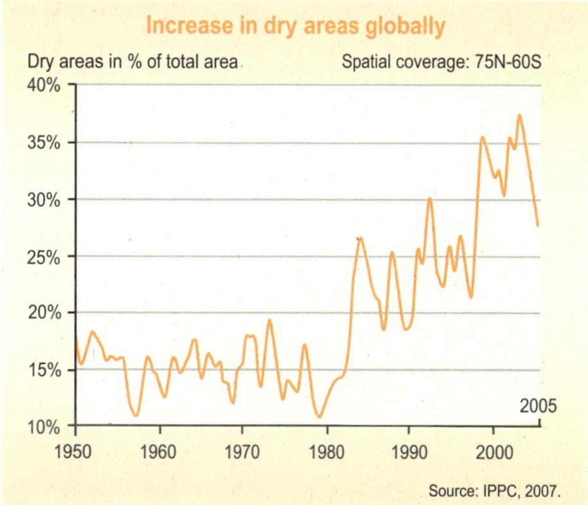
Increase in dry areas globally
Dry areas in % of total area
Spatial coverage: 75N-60S
40%
35%
30%
25%
20%
15%
10%
2005
1950
1960
1970
1980
1990
2000
Source: IPPC, 2007.

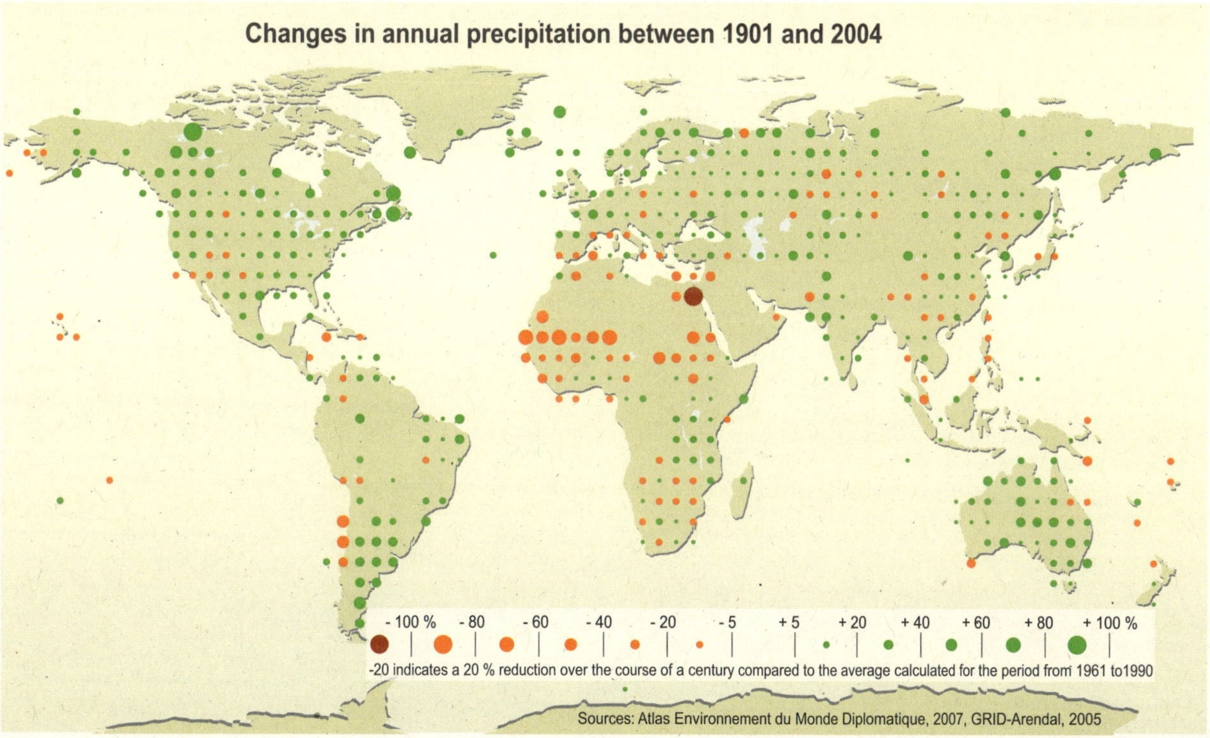
Changes in annual precipitation between 1901 and 2004
- 100 %
- 80
- 60
- 40
- 20
- 5
+ 5
+ 20
+ 40
+ 60
+ 80
+ 100 %
-20 indicates a 20 % reduction over the course of a century compared to the average calculated for the period from 1961 to1990
Sources: Atlas Environnement du Monde Diplomatique, 2007, GRID-Arendal, 2005

The graphic below shows that while natural disasters reported have increased globally, some of this increase can be attributed to the increase in communications: while disasters happened before, few people would have known about them. But the distinction between climate-dependent disasters like tropical cyclones and others not influenced by climate (such as earthquakes) shows a clear difference in trend: while the frequency of earthquakes shows little variability, floods and cyclones have occurred more frequently in the last 30 years. However, while intense tropical cyclones have increased since about 1970, high variability over several decades and a lack of systematic high quality observation before satellite observations make it difficult to detect long-term trends.

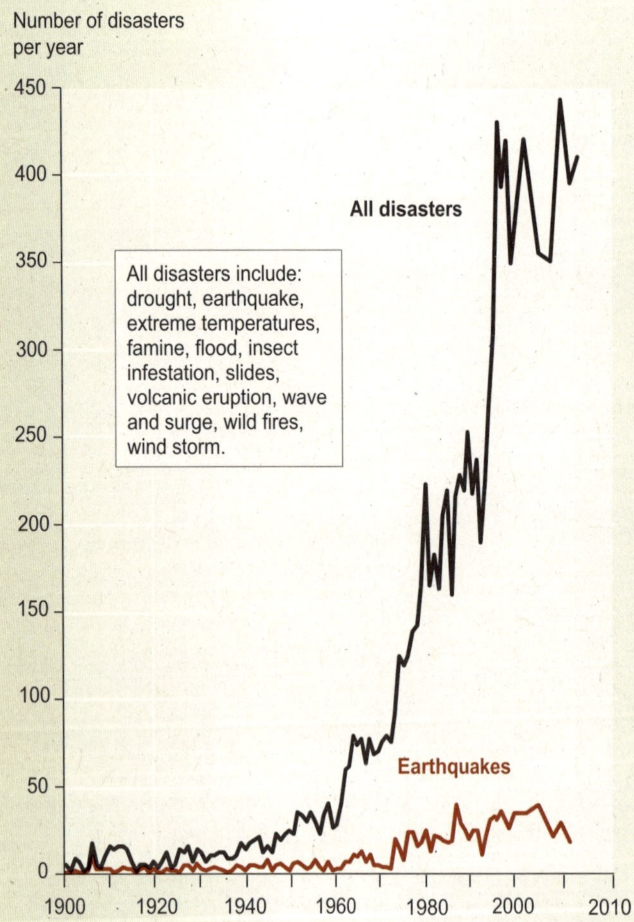

Trends in number of reported disasters

Much of the increase in the number of hazardous events reported is probably due to significant improvements in information access and also to population growth, but the number of floods and cyclones reported is still rising compared to earthquakes. Is global warming affecting the frequency of natural hazards?

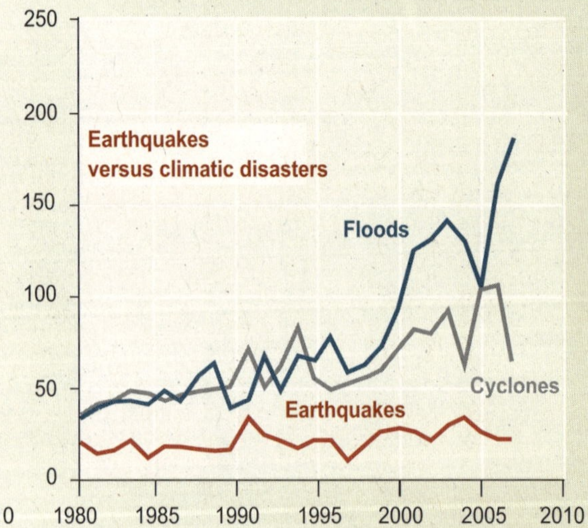

Source: CRED Annual Disaster Statistical Review 2006, 2007.

Reported changes in physical and biological systems between 1970 and 2004

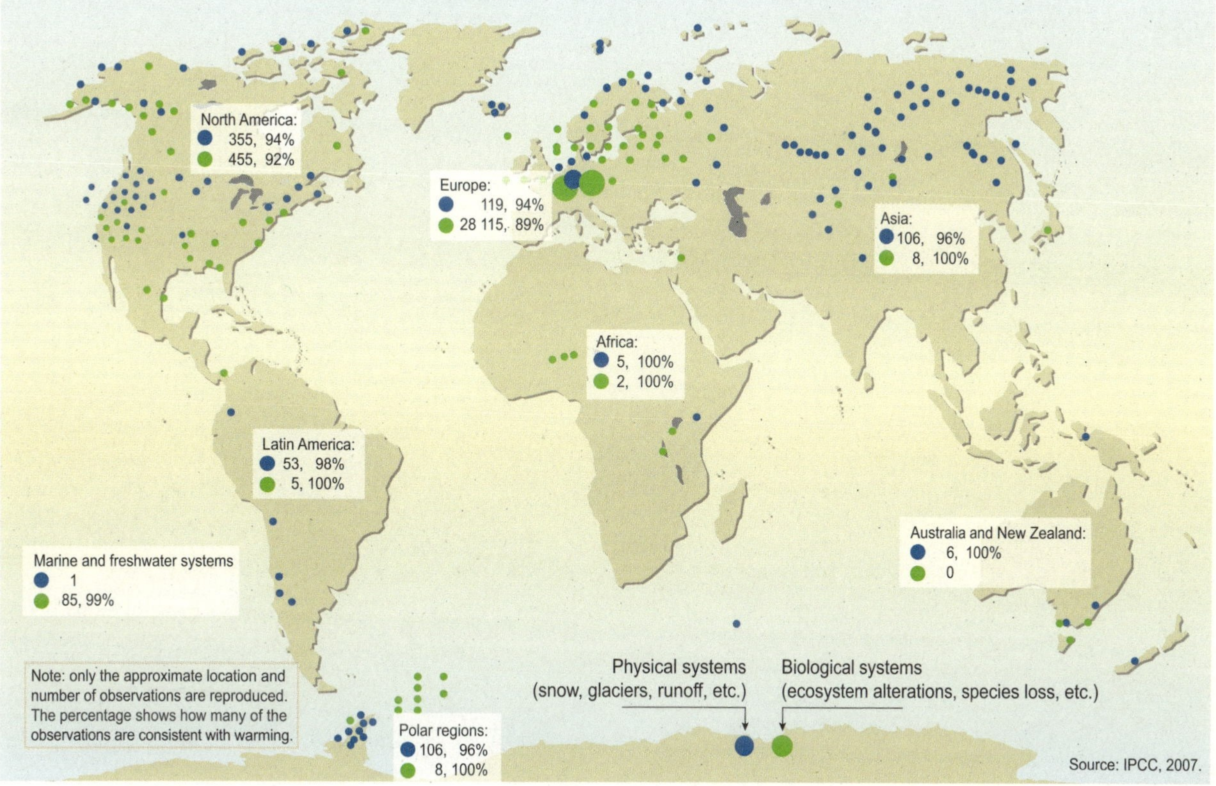

Natural systems affected

Observations from across the world show that many natural systems are being affected by regional climate changes, especially by temperature increases.

Other effects of regional climate change on humans and ecosystems than the ones already described are emerging. Many are difficult to pinpoint, both because of steps to adapt to changing climate, and because of factors unrelated to it. They range across areas as different as planting crops earlier in spring to changes in allergic pollen distribution in the Northern Hemisphere, changes in the extension of areas where infectious diseases are transmitted or effects on activities depending for example on snow and ice cover such as mountain sports.

Causes of change

Carbon cycle

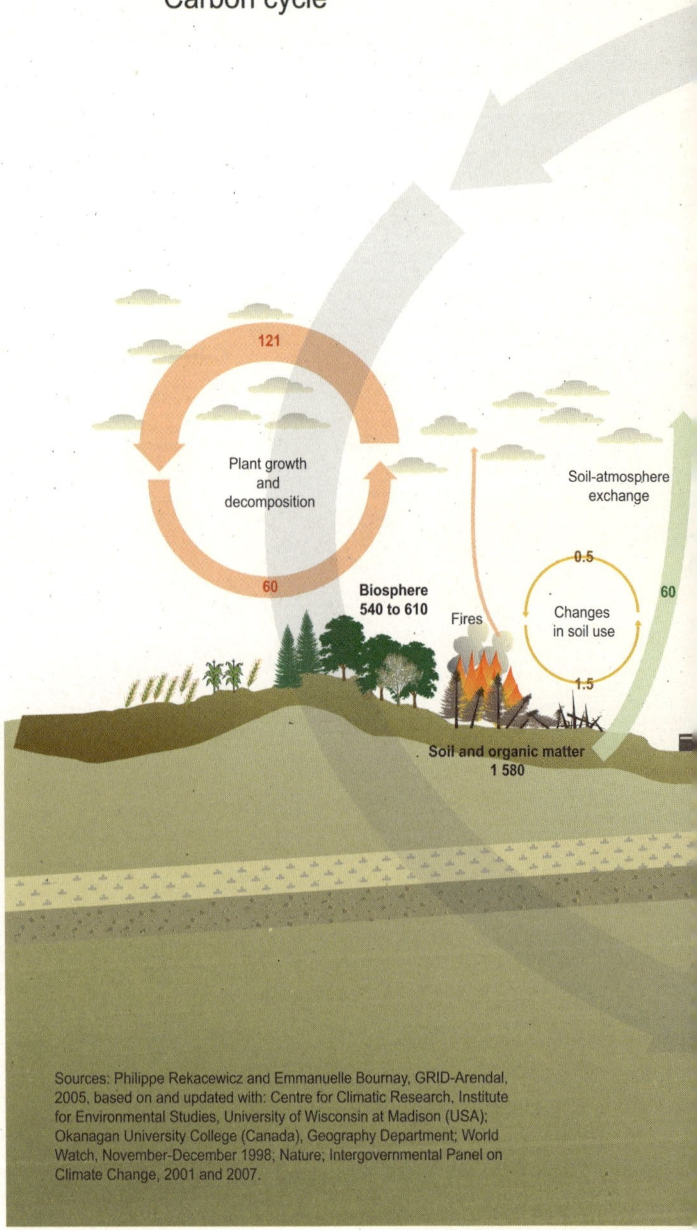

Sources: Philippe Rekacewicz and Emmanuelle Bournay, GRID-Arendal, 2005, based on and updated with: Centre for Climatic Research, Institute for Environmental Studies, University of Wisconsin at Madison (USA); Okanagan University College (Canada), Geography Department; World Watch, November-December 1998; Nature; Intergovernmental Panel on Climate Change, 2001 and 2007.

There is no longer much doubt that most of the observed increase in global average temperatures since the mid-20th century is due to the observed increase in greenhouse gases (GHGs) emanating from human activities. It is likely that there has been significant human-caused warming over the past 50 years, averaged over each continent except Antarctica. Over that period the combined effect of natural variations in solar radiation and volcanic eruptions would have produced *cooler* temperatures, not warmer. There has in fact been a cooling influence on the atmosphere, caused by *aerosols*, some of them caused naturally, for instance by volcanic eruptions, and some by human activities, principally emissions of sulphate, organic and black carbon, nitrate and dust. These aerosols reflect some of the Sun's rays back to space or absorb some of them, in either case preventing them from reaching the surface of the Earth.

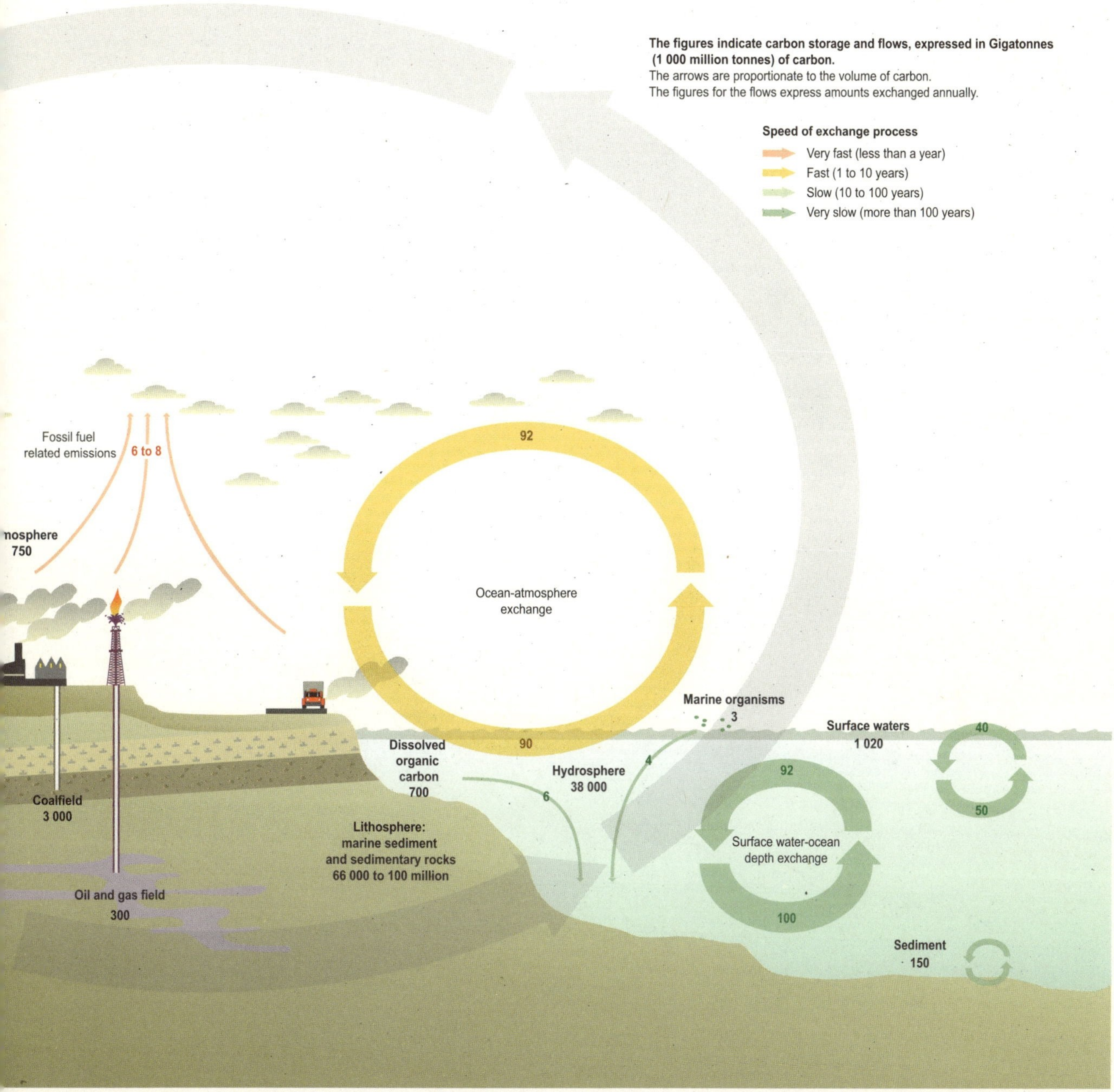

The figures indicate carbon storage and flows, expressed in Gigatonnes (1 000 million tonnes) of carbon.
The arrows are proportionate to the volume of carbon.
The figures for the flows express amounts exchanged annually.
Speed of exchange process
Very fast (less than a year)
Fast (1 to 10 years)
Slow (10 to 100 years)
Very slow (more than 100 years)
Fossil fuel related emissions
6 to 8
nosphere
750
92
Ocean-atmosphere exchange
90
Marine organisms
3
Surface waters
1 020
40
50
Dissolved organic carbon
700
6
Hydrosphere
38 000
4
92
Surface water-ocean depth exchange
100
Coalfield
3 000
Lithosphere: marine sediment and sedimentary rocks
66 000 to 100 million
Oil and gas field
300
Sediment
150

Human activity

By comparing a number of natural factors and human activities, and their effect on climate, scientists have proved that human activity is responsible for part of the temperature rise. So computer models which include the human influence on the climate show a clear increase in temperature. This accurately reflects the actual pattern of warming we have been experiencing. In contrast, temperatures predicted by models that take into account natural factors alone stay well below the actual temperatures measured.

Not only average temperatures but also other aspects of the climate are changing because of distinguishable human influences. Human activities have contributed to sea level rise during the second half of the 20th century; they have probably helped to change wind patterns and increased temperatures on extreme hot nights, cold nights and cold days. In addition, our actions may have contributed to increasing the risk of heat waves, the area affected by drought since the 1970s, and the frequency of episodes of heavy precipitation. Human-caused warming over the last three decades has had a discernible global influence on the changes observed in many physical and biological systems.

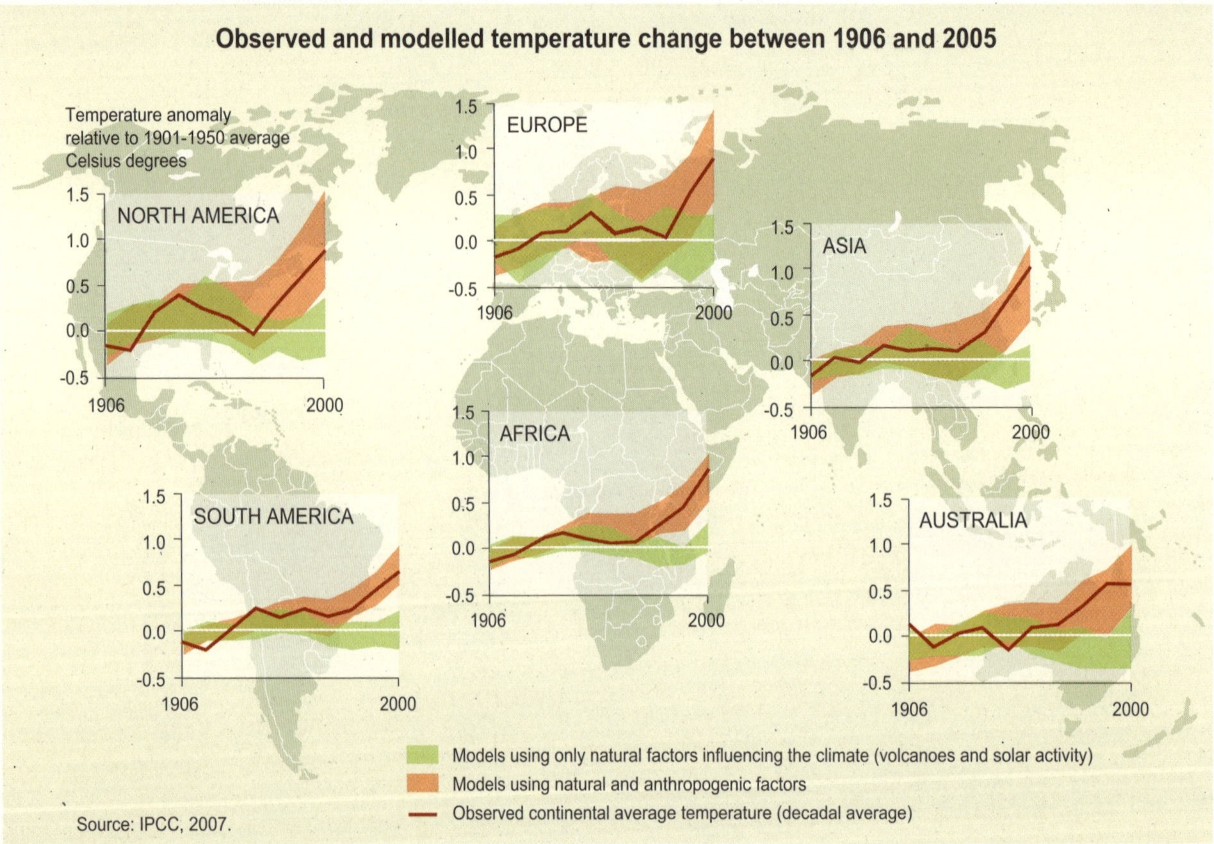

Climate change global processes and effects

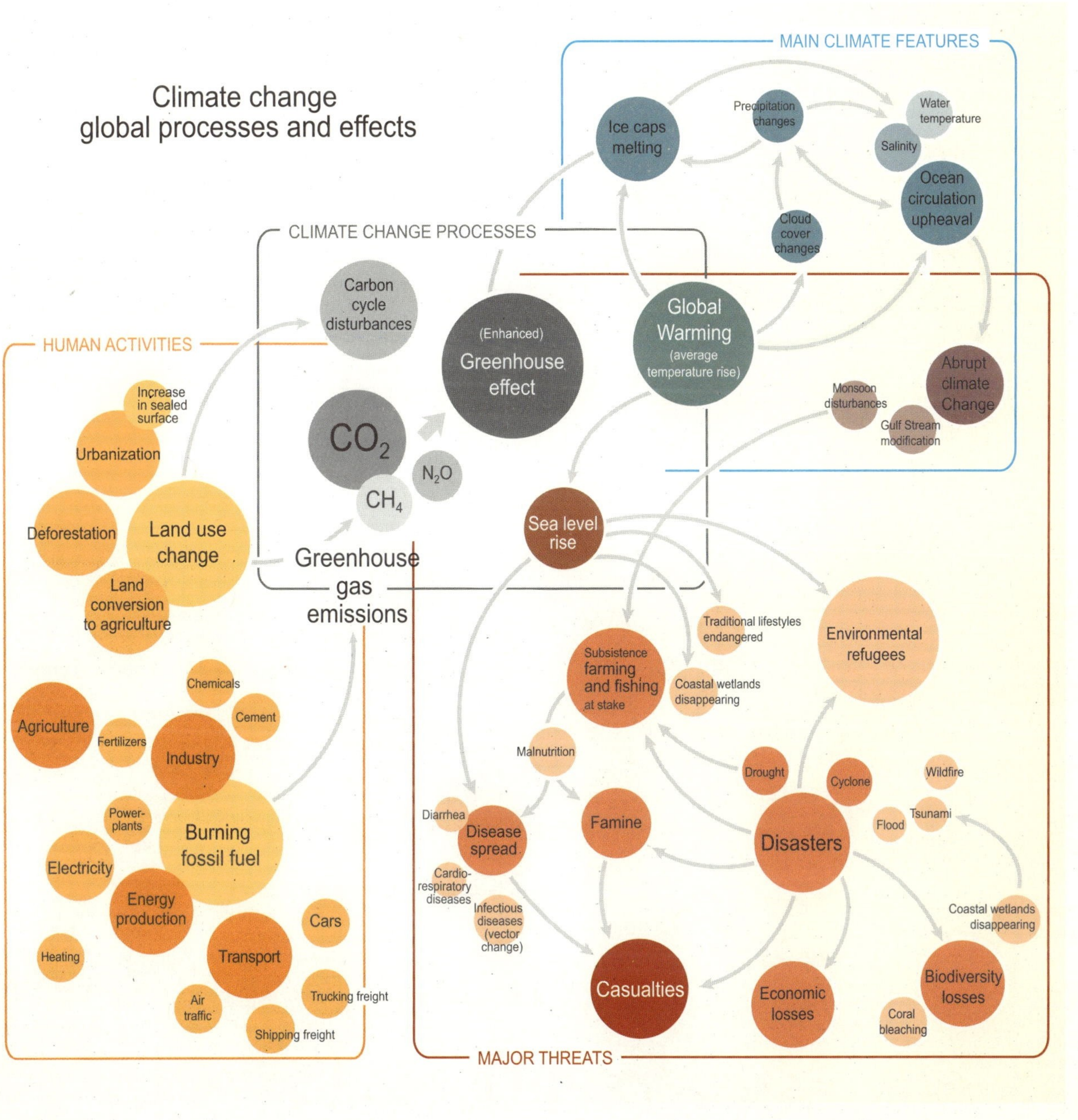

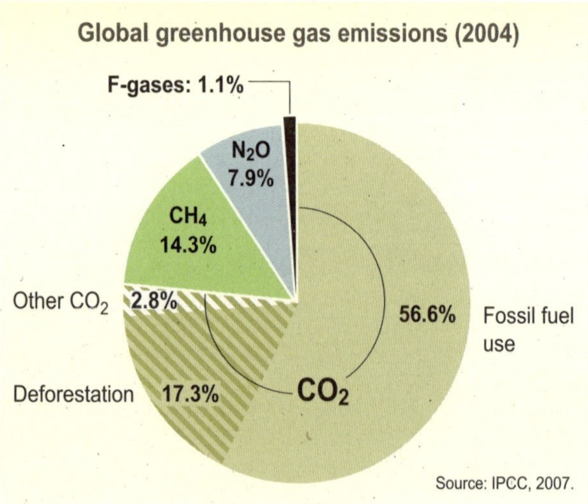

Source: IPCC, 2007.

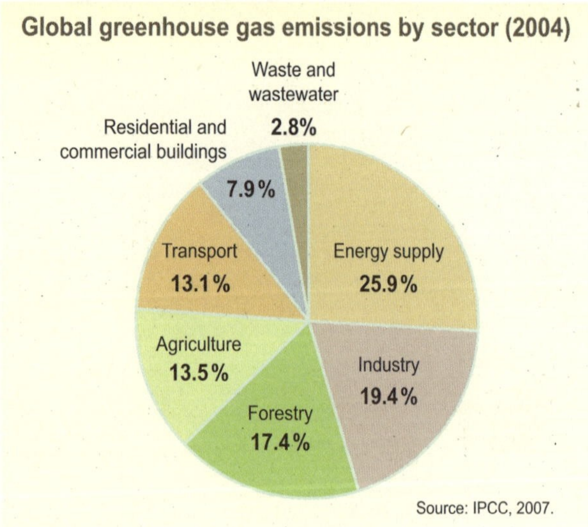

Source: IPCC, 2007.

Greenhouse gases

The Kyoto Protocol, the international climate change agreement[1] lists six greenhouse gases (or groups of gases) whose emissions signatories to the Protocol agree to reduce. There are also other GHGs apart from the ones covered by the Protocol. But these six gases/groups of gases make up a big chunk of overall GHG emissions from human activities and are the most relevant in terms of immediate human responsibility:

- carbon dioxide (CO_2)
- methane (CH_4)
- nitrous oxide (N_2O)
- sulphur hexafluoride (SF_6)
- hydrofluorocarbons (HFCs)
- perfluorocarbons (PFCs)

The last three (SF_6, HFCs and PFCs) are sometimes known collectively as F-gases.

Global greenhouse gas emissions from human activities have grown since pre-industrial times, with an increase of 70 per cent between 1970 and 2004. Global atmospheric concentrations of carbon dioxide (CO_2), methane (CH_4) and nitrous oxide (N_2O) have risen markedly because of human activities since 1750 – the year

1. The Kyoto Protocol sets the rules and procedures needed to achieve the ultimate objective of the Convention, which is: "(...) to achieve, (...), stabilization of greenhouse gas concentrations in the atmosphere at a level that would prevent dangerous anthropogenic interference with the climate system." Such a level should be achieved within a time frame sufficient to allow ecosystems to adapt naturally to climate change, to ensure that food production is not threatened and to enable economic development to proceed in a sustainable manner.

Main greenhouse gases

Gas name	Pre-industrial concentration (ppmv *)	Concentration in 1998 (ppmv)	Atmospheric lifetime (years)	Main human activity source	GWP **
Water vapour	1 to 3	1 to 3	a few days	-	-
Carbon dioxide (CO_2)	280	365	variable	Fossil fuels, cement production, land use change	1
Methane (CH_4)	0.7	1,75	12	Fossil fuels, rice paddies waste dumps, livestock	21
Nitrous oxide (N_2O)	0.27	0,31	114	Fertilizers, combustion industrial processes	310
HFC 23 (CHF_3)	0	0,000014	250	Electronics, refrigerants	12 000
HFC 134 a (CF_3CH_2F)	0	0,0000075	13.8	Refrigerants	1 300
HFC 152 a (CH_3CHF_2)	0	0,0000005	1.4	Industrial processes	120
Perfluoromethane (CF_4)	0,0004	0,00008	>50 000	Aluminium production	5 700
Perfluoroethane (C_2F_6)	0	0,000003	10 000	Aluminium production	11 900
Sulphur hexafluoride (SF_6)	0	0,0000042	3 200	Dielectric fluid	22 200

* ppmv = parts per million by volume

** GWP = Global warming potential (for 100 year time horizon).

World greenhouse gas emissions by sector

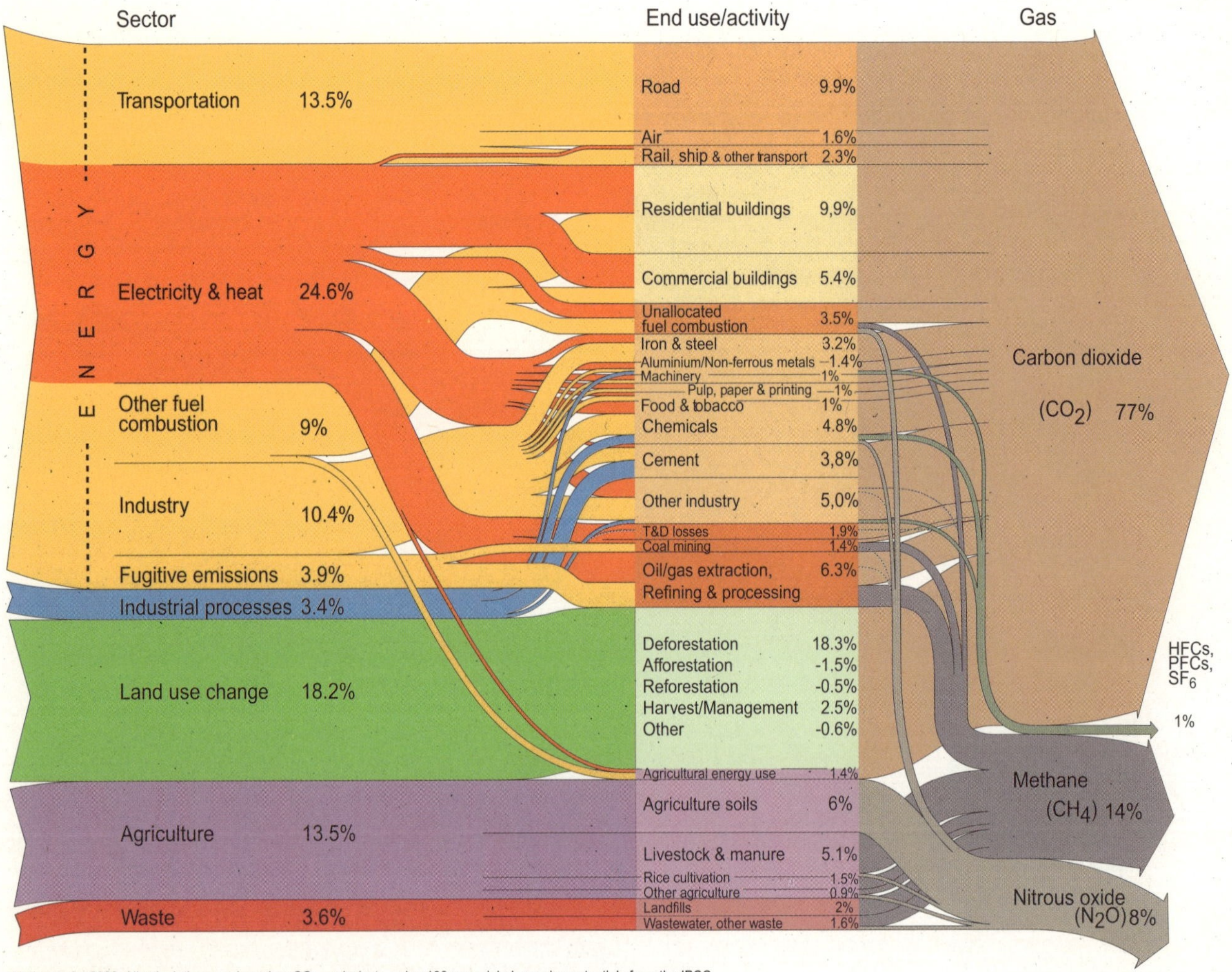

All data is for 2000. All calculations are based on CO_2 equivalents, using 100-year global warming potentials from the IPCC (1996), based on a total global estimate of 41 755 $MtCO_2$ equivalent. Land use change includes both emissions and absorptions. Dotted lines represent flows of less than 0.1% percent of total GHG emissions.

Source: World Resources Institute, Climate Analysis Indicator Tool (CAIT), Navigating the Numbers: Greenhouse Gas Data and International Climate Policy, December 2005; Intergovernmental Panel on Climate Change, 1996 (data for 2000).

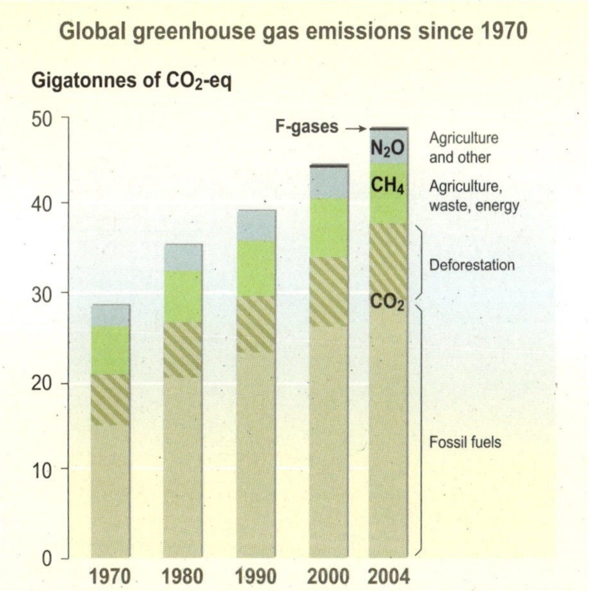

Source: IPCC, 2007.

that commonly marks the beginning of industrial activities – and are now far higher than pre-industrial levels, as shown by ice cores spanning many thousands of years. Levels of the first two gases are far above their natural range over the last 650,000 years. The global atmospheric concentration of CO_2 before the start of the Industrial Revolution was around 280 parts per million (ppm). By 2005 it had reached 379 ppm. And the increase is accelerating: The annual growth rate was larger from 1995 to 2005 than at any time since continuous atmospheric measurements began in the 1950s.

Global Warming Potential (GWP) and CO_2 equivalence

Each of the greenhouse gases affects the atmosphere to a different degree, and survives for a different length of time. The extent to which a given GHG is estimated to contribute to global warming is described as its *Global Warming Potential* (GWP). To make the effects of different gases comparable, the GWP expresses the factor by which the gas in question is more (or less) damaging than the same mass of CO_2 over a given period of time, so the GWP of CO_2 is always 1. Gases which cause much more warming than CO_2 may in turn decay faster than it does, so they may pose a considerable problem for a few years but a smaller one later. Equally, others may decay slower and pose a greater problem over a long period of time.

To talk of CO_2-equivalent emissions is to take carbon dioxide as a benchmark and to describe the amount of CO_2 that would cause the same amount of warming over a specified timespan as would be caused by one of the other GHGs.

For example, the GWP for methane over 100 years is 25 and for nitrous oxide 298. This means that emissions of one metric tonne of methane or of nitrous oxide are equivalent to emissions of 25 and 298 metric tonnes of carbon dioxide respectively. One of the F-gases, HFC_{23}, is 12,000 times more potent than CO_2 over 20 years, becoming even more potent (and thus "dangerous" for the climate) over 100 years, the period in which its GWP reaches 14,800.

Actual and projected energy demand

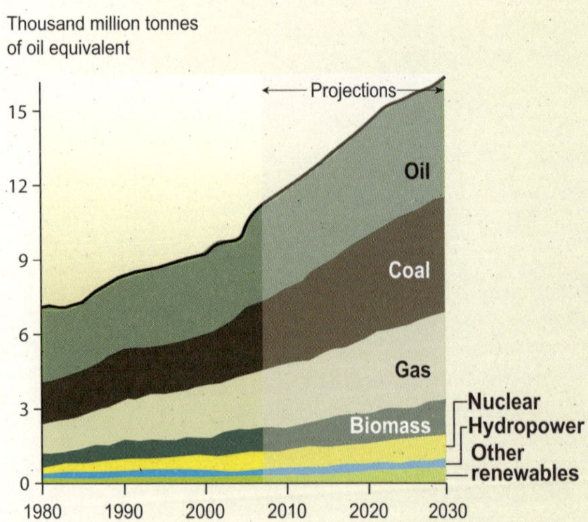

Note: All statistics refer to energy in its original form (such as coal) before being transformed into more convenient energy (such as electrical energy).

Source: International Energy Agency (IEA), World Energy Outlook 2008.

Feedbacks

One factor which complicates climate science – and therefore leads to wide ranges of uncertainty – is the existence of feedbacks. These are interactions between different parts of the climate system, which can mean a process or event sets off changes which in their turn influence the initial trigger. One example is the reduction of ice and snow, both on land and at sea. Ice, being white, reflects up to 90 per cent of the Sun's radiation reaching its surface back out into space, preventing it from intensifying atmospheric warming. But when it melts it may expose earth, vegetation, rock or water, all of which are darker in colour and therefore more likely to absorb radiation instead of reflecting it. So the initial melting can cause a feedback which helps to quicken its pace. Another possible feedback is the thawing of the permafrost in high northern latitudes. As it melts, it could release big quantities of carbon dioxide and methane which at the moment are retained below the frozen soil layer. If that happened, it would accelerate the warming already under way. Another expected feedback: Higher temperatures of both land and ocean have the tendency to reduce their uptake of atmospheric carbon dioxide, increasing the amount of CO_2 that remains in the atmosphere. These are all positive feedbacks because they intensify the original process. Negative feedbacks, on the other hand, are changes in the environment that lead to a compensating process and mitigate the change itself.

Feedbacks in Climate Processes

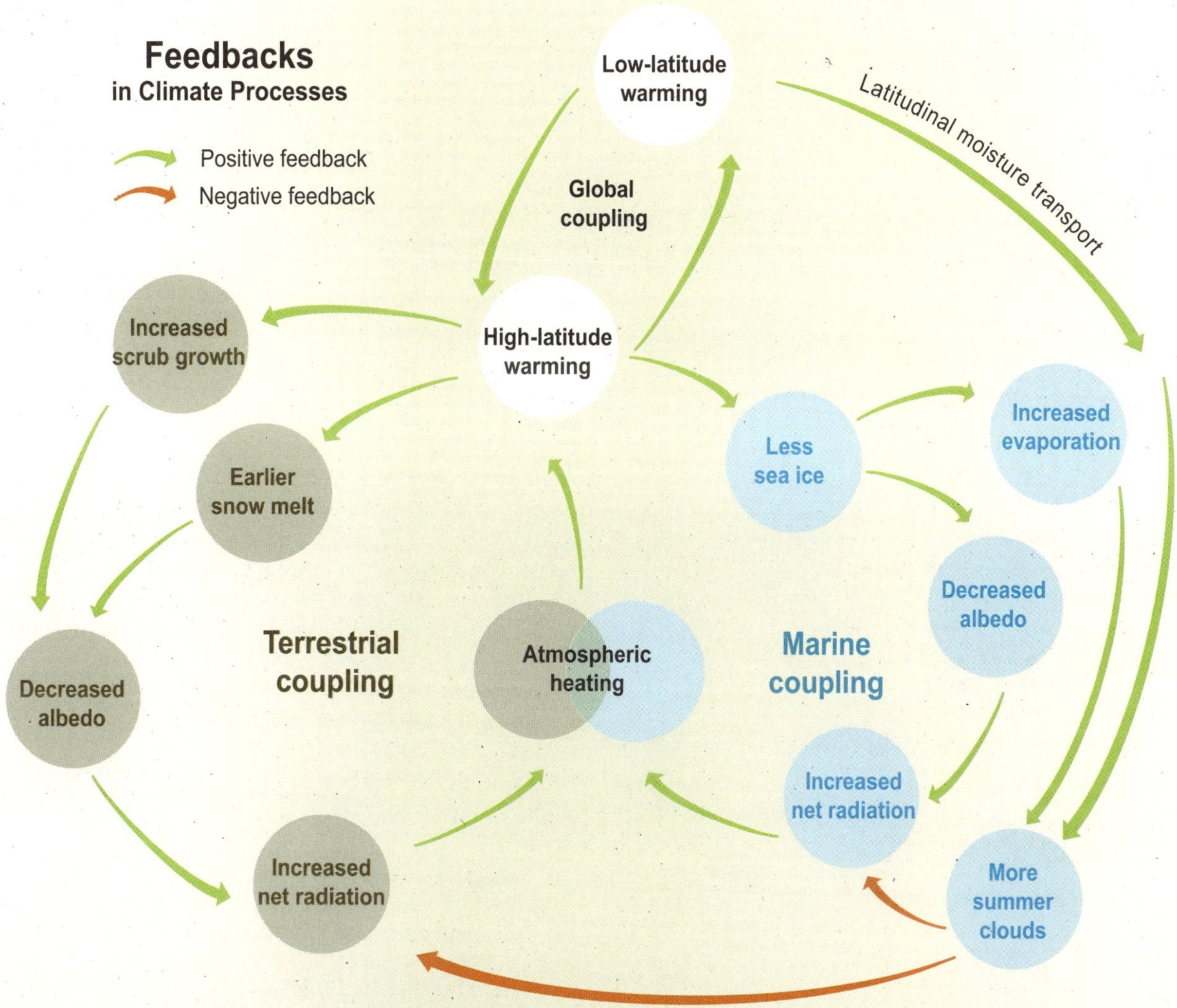

Positive feedbacks intensify the original process.
Negative feedbacks lead to a compensating process.

Source: Hugo Ahlenius/ GRID-Arendal, *Global Outlook for Ice and Snow*, 2008.

Projected climate change and its impacts

The previous sections have dealt with changes and effects happening today. The rest of the guide concentrates on what is yet to come.

If current policies to mitigate climate change and related steps towards sustainable development remain unchanged, global GHG emissions will continue to grow over the next few decades. The growth will not be modest, either: The IPCC projects an increase in global GHG emissions from 25 to 90 per cent between 2000 and 2030. (There is a wide margin of uncertainty because of the very different assumptions made in each socio-economic scenario considered by the IPCC). It expects the fossil fuels, oil, coal and gas, to continue to dominate the energy mix till beyond 2030, regardless of the scenario.

Continued GHG emissions at or above current rates will cause further warming and induce many changes in the global climate system during this century that would be larger than those observed during the 20th century.

Today's emissions influence the atmosphere for years to come

Even if GHG and aerosol concentrations were kept constant at today's levels (2000), some anthropogenic warming and sea level rise would continue for many centuries. The climate reacts over long periods to influences upon it; many GHGs remain in the atmosphere for thousands of years.

Backed up by new studies and observations, the IPCC is more certain of the accuracy of the projected warming patterns and other regional climatic effects than it was in the Third Assessment Report. These include wind-pattern changes, precipitation, and some changes in weather extremes and sea ice.

Regional-scale changes include:

- most warming over land and at the highest northern latitudes, and least over the Southern Ocean and parts of the North Atlantic;
- contraction of the area covered by snow, increases in the depth at which most permafrost will thaw, and a decrease in the extent of sea ice;
- increase in the frequency of extremes of heat, heat waves and heavy precipitation;
- a likely increase in tropical cyclone intensity;
- a shift towards the poles of storms outside the tropics;
- increases in precipitation in high latitudes, and likely decreases in most sub-tropical land regions.

The IPCC scenarios (often referred to as the SRES scenarios, for "Special Report on Emissions Scenarios" published by the IPCC in 2000) explore alternative development pathways. They take into account demographic, economic and technological factors and their resulting GHG emissions. The emissions projections based on those different assumptions are widely used in forecasting future climate change, vulnerability and impacts. It is open to anyone to decide which of the different scenarios seems most probable, as the IPCC doesn't take the risk of attaching any probability to any of them.

Scenarios

A scenario is a plausible and often simplified description of how the future may develop, based on a coherent set of assumptions: a set of working hypotheses about how society may develop, and what this will mean for the climate.

The A1 scenario describes a future world of very rapid economic growth, a global population that peaks in mid-century and declines thereafter, and the rapid introduction of new and efficient technologies. Specific regional patterns tend to disappear as a result of increased cultural interaction. The gap in per capita income between regions is substantially reduced. A1 develops into three groups that describe alternative developments of energy supply: fossil intensive (A1FI), non-fossil energy sources, or a balance between all sources (A1B).

The A2 scenario describes a very heterogenous world, based on the preservation of local identities. Fertility patterns across regions converge slowly, resulting in a continuously increasing population. Economic development is regionally oriented, and per capita economic growth and technological change more fragmented than in A1.

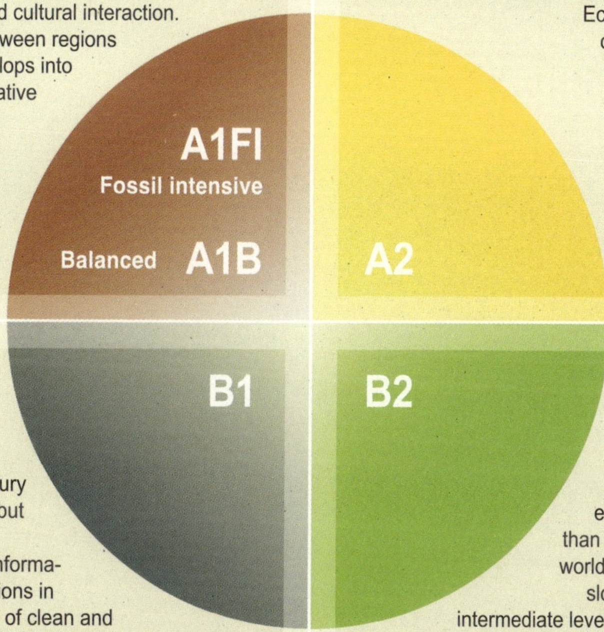

The B1 scenario describes a convergent world with a global population that peaks in mid-century and declines thereafter as in A1, but with rapid change in economic structures toward a service and information economy. It describes reductions in consumption and the introduction of clean and resource-efficient technologies. The emphasis is on global solutions to sustainability, including improved equity, but without additional climate initiatives.

The B2 scenario describes a world with an emphasis on local solutions to economic social and environmental sustainability rather than the global approach in B1. It is a world with a population increasing at a slower rate than in other scenarios, intermediate levels of economic development, and slow but diverse technological change. Society is oriented towards environmental protection and social equity and focuses on the local and regional level.

Source: GRID-Arendal based on information from IPCC, 2001.

Variations in the Earth's surface temperature: year 1000 to 2100

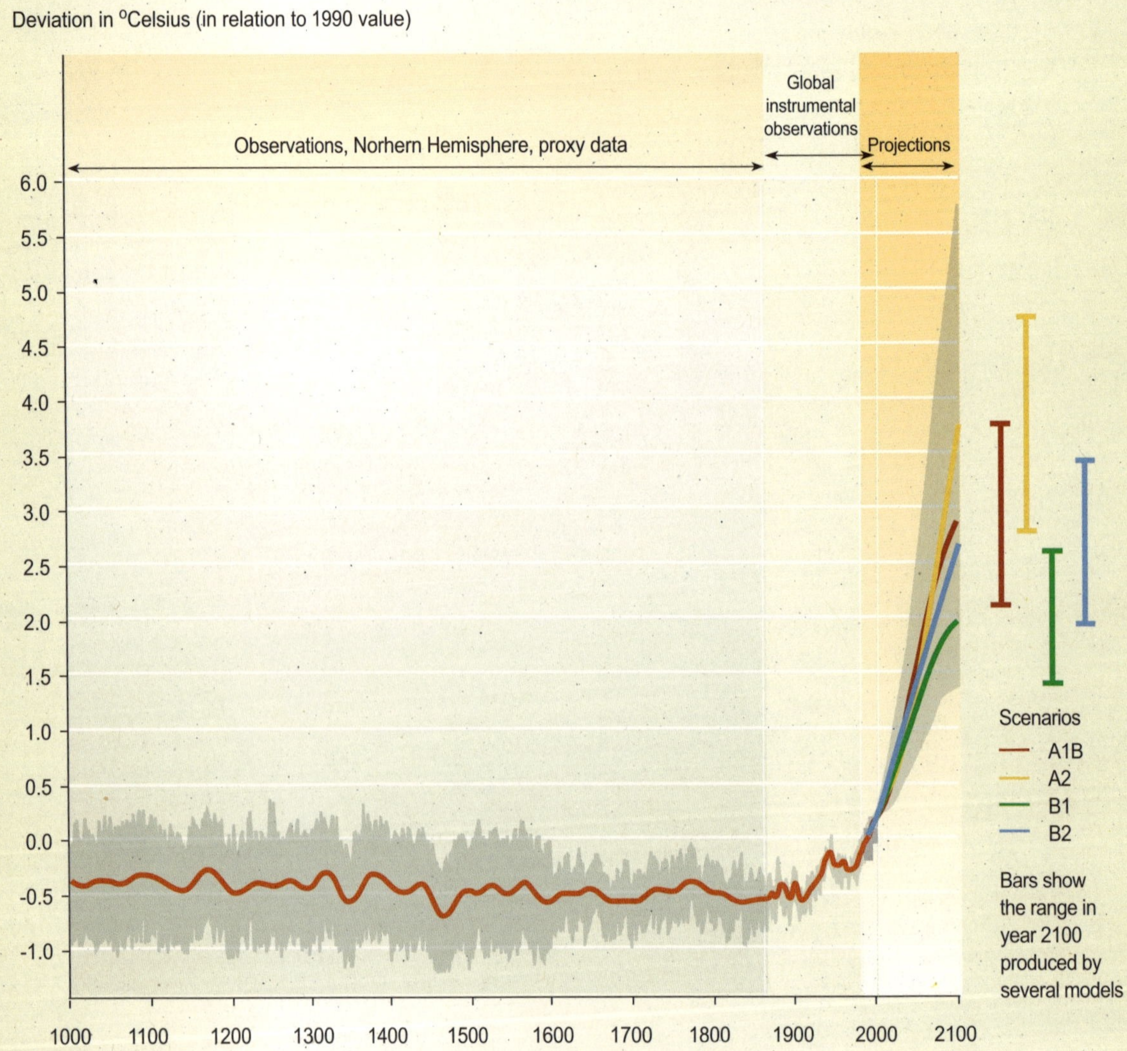

Source: UNEP&GRID/Arendal, Vital Climate Graphics update, 2005.

Sea level rise

Because our understanding of some important causes of sea level rise is too limited, the IPCC is careful in making predictions: it does not provide a best estimate or an upper limit of the rise to be expected. The range of rise projected for 2090–2099 is from 18 to 59 cm relative to 1980–1999, but this does not include the possible acceleration of ice flow from the polar ice sheets and is therefore not an upper boundary. Other factors contributing to uncertainty in this area include feedbacks between the climate and the *carbon cycle* and the expansion of ocean water due to warming ("thermal expansion").

The shrinking of the Greenland ice sheet is projected to contribute to rising sea levels until after 2100. Models suggest a nearly complete melting of the ice sheet.

The Antarctic ice sheet is projected to remain too cold for widespread surface melting: instead, it is expected to grow because of increased snowfall. But a net loss of ice from Antarctica could still be possible, depending on the extent and pace at which ice moves from the land into the sea.

Ecosystems[2]

Since the IPCC's Third Assessment Report, confidence has increased that a 1–2°C increase in global mean temperature above 1990 levels (about 1.5–2.5°C above pre-industrial levels) poses significant risks to many unique and threatened systems, including many biodiversity hotspots.

2. An ecosystem is a natural unit consisting of all the plants, animals and micro-organisms in a defined area functioning together with all the non-living physical factors of the environment.

These ecosystem changes will be accompanied by shifts in the geographical ranges of both animal and plant species, with mainly harmful consequences for the natural world and for the goods and services which ecosystems provide – like water and food. Many ecosystems are "likely" to exhaust their capacities to cope with changes inflicted on them by climate change and the upheavals associated with it.

Ecosystems will probably find their net carbon uptake peak before the middle of the century and then weaken or even reverse, which would amplify climate change (a positive *feedback*).

Approximately 20–30 per cent of species are at increased risk of extinction if global average warming exceeds 1.5–2.5°C. As global average temperature increase exceeds about 3.5°C, model projections suggest significant extinctions of 40–70 per cent of known species around the globe. This is one of the irreversible impacts of climate change.[3]

Food security and human health

The effects of more frequent and intense extreme weather events will cause emergencies and reverse progress towards development. Extreme events coupled with sea level rise, are expected to be mainly adverse for humans. Food is an obvious worry. In higher latitudes there may be an initial slight increase in crop productivity for temperature rises below 3°C, to be followed by a decrease in some areas. For lower latitudes, productivity may decrease for even small temperature rises.

3. Many species have evolved to live in a particular and often narrowly-defined environment. As temperatures rise and other effects of a changing climate intensify, their environments are likely to change too quickly for them to either adapt to or to migrate to somewhere more suitable.

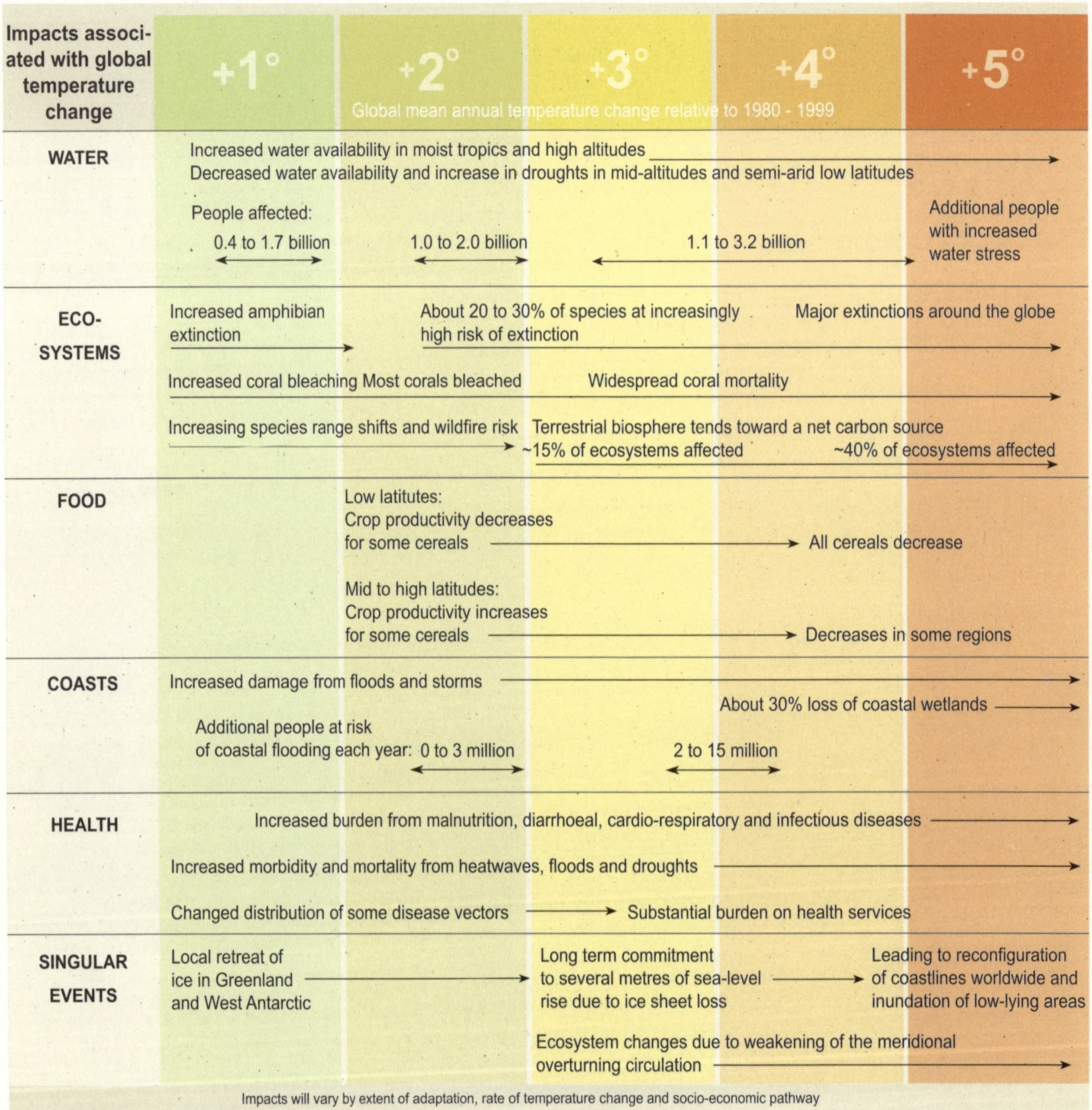
Impacts associated with global temperature change
+1°
+2°
+3°
+4°
+5°
Global mean annual temperature change relative to 1980 - 1999
WATER
Increased water availability in moist tropics and high altitudes
Decreased water availability and increase in droughts in mid-altitudes and semi-arid low latitudes
People affected:
0.4 to 1.7 billion
1.0 to 2.0 billion
1.1 to 3.2 billion
Additional people with increased water stress
ECO-SYSTEMS
Increased amphibian extinction
About 20 to 30% of species at increasingly high risk of extinction
Major extinctions around the globe
Increased coral bleaching
Most corals bleached
Widespread coral mortality
Increasing species range shifts and wildfire risk
Terrestrial biosphere tends toward a net carbon source
~15% of ecosystems affected
~40% of ecosystems affected
FOOD
Low latitutes:
Crop productivity decreases for some cereals
All cereals decrease
Mid to high latitudes:
Crop productivity increases for some cereals
Decreases in some regions
COASTS
Increased damage from floods and storms
About 30% loss of coastal wetlands
Additional people at risk of coastal flooding each year:
0 to 3 million
2 to 15 million
HEALTH
Increased burden from malnutrition, diarrhoeal, cardio-respiratory and infectious diseases
Increased morbidity and mortality from heatwaves, floods and droughts
Changed distribution of some disease vectors
Substantial burden on health services
SINGULAR EVENTS
Local retreat of ice in Greenland and West Antarctic
Long term commitment to several metres of sea-level rise due to ice sheet loss
Leading to reconfiguration of coastlines worldwide and inundation of low-lying areas
Ecosystem changes due to weakening of the meridional overturning circulation
Impacts will vary by extent of adaptation, rate of temperature change and socio-economic pathway

The health of millions of people could be at risk from increases in malnutrition, extreme weather, more diarrhoeal diseases, heart and breathing problems caused by climate-induced ground-level ozone, and the spread of some infectious diseases. But there may be some benefits, for instance to those who suffer from the effects of very cold weather. For other issues, effects will be mixed. In some places the geographical range of malaria distribution could contract, while elsewhere it will expand and the transmission season may be changed. Overall, the negative health effects of rising temperatures are expected to outweigh the benefits, especially in developing countries.

Freshwater supply

Climate change is expected to have a crucial impact on all sectors and regions. It would worsen current water **stress**[4] caused by population growth and economic and land-use change.

Regionally, glaciers and snow fields are crucial sources of fresh water. They have undergone recent widespread and severe loss from melting, and this is projected to accelerate this century, reducing water

4. One definition of water stress is the situation that occurs when a country uses more than 20 per cent of its renewable water supply.

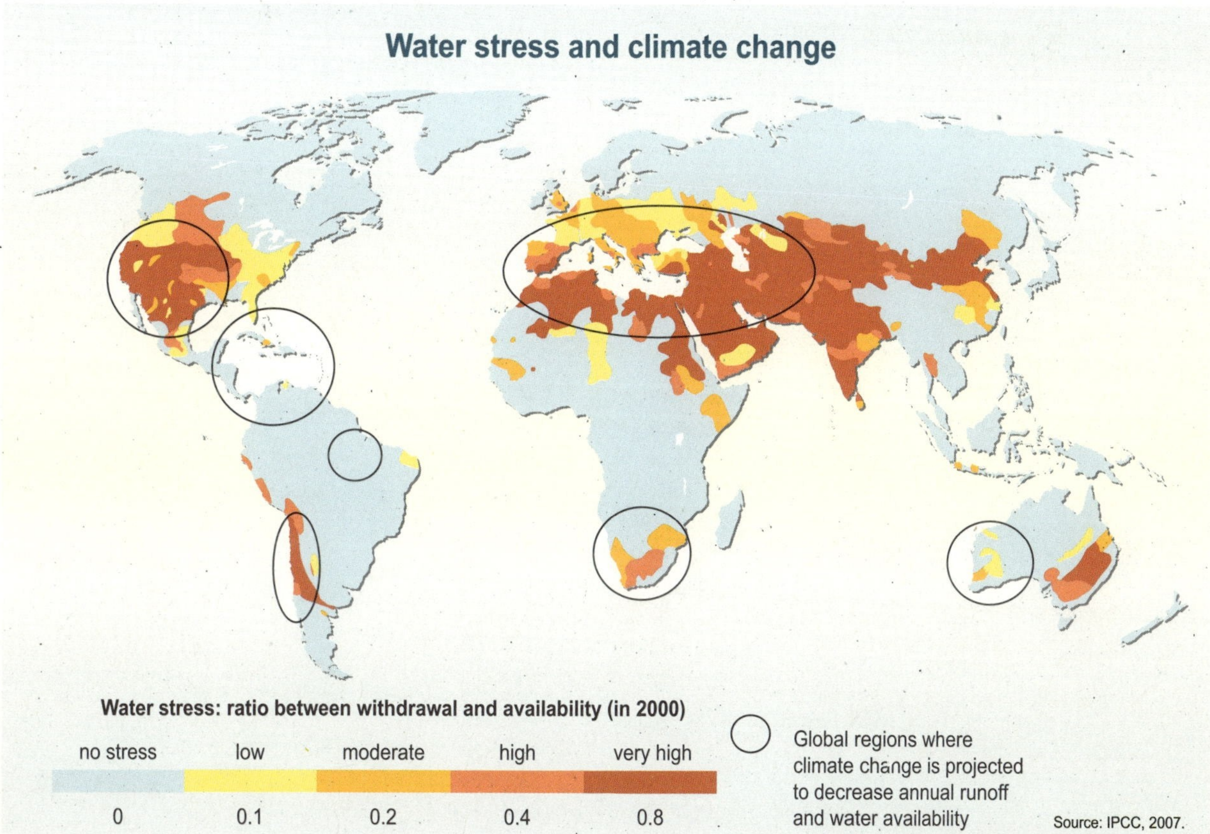

availability and the potential for hydropower (often a feasible alternative to fossil fuel-based electricity generation which reduces CO_2 emissions). Climate change is also expected to change the seasonal flows in regions fed by melt water from mountain ranges, like the Hindu Kush, the Himalayas and the inter-tropical Andes. More than a sixth of the world's population lives in these regions. Two thousand million people depend on the water provided by seven of the major rivers in Asia, all of them originating in the Himalayas.

Changes in precipitation and temperature also affect run-off and water availability. Run-off[5] will increase by 10–40 per cent by mid-century at higher latitudes and in some wet tropical areas, and decrease by 10–30 per cent over some dry mid-latitude and tropical regions. Some semi-arid areas, for example around the Mediterranean, in the western USA, southern Africa and north-eastern Brazil, will have less water. Areas affected by drought are projected to increase, threatening food, water, energy production and health by increasing malnutrition and infectious and respiratory diseases. Large regional increases in demand for irrigation water are projected.

The negative impacts on freshwater systems will outweigh the benefits of climate change. Available research suggests a future increase in heavy rainfall events in many regions and some regions where **average** rainfall is projected to decrease. This will mean an increased flood risk. It is likely that up to 20 per cent of the world's people will be living in areas where the river flood potential could increase by the 2080s. More frequent and severe floods and droughts will harm sustainable development, rising temperatures will affect fresh water quality, and in coastal areas rising sea levels will mean more saline contamination of groundwater.

5. The water that falls to Earth and does not infiltrate the ground or evaporate but flows over or through the ground to swell surface or ground water sources.

Ocean acidification

The increased amount of carbon dioxide from human activities that has entered the oceans via the atmosphere since about 1750 has made them more acidic, with an average decrease in pH (the measure of the acidity or alkalinity of a solution) of 0.1 units. Increasing atmospheric CO_2 concentrations are causing further acidification. Today the average ocean surface pH is about 8.1. Projections suggest a further acidification over this century, leading to a reduction in average global surface ocean pH of between 0.14 and 0.35 units. This progressive acidification is expected to harm marine creatures which form shells, for instance corals, and the species which depend upon them.

Studies published since the 2001 IPCC assessment report allow a more systematic understanding of the timing and extent of impacts linked to different rates of climate change.

Global ocean acidification

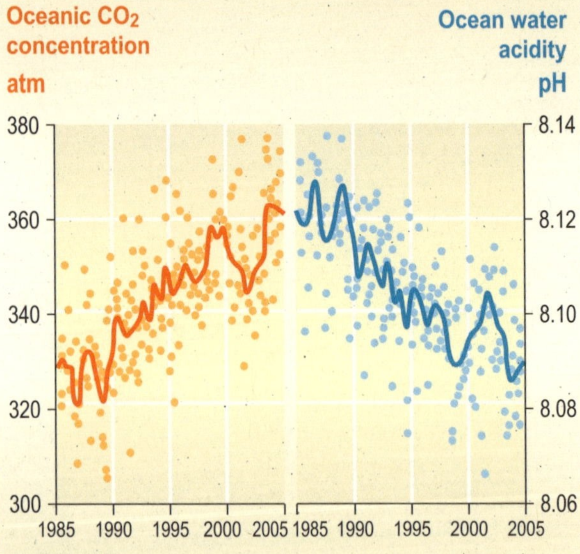

Source: IPCC, 2007.

Examples of major projected impacts on selected sectors

Climate driven phenomena	Agriculture, forestry and ecosystems	Water resources	Human health	Industry, settlements and society
TEMPERATURE CHANGE Over most land areas, warmer and fewer cold days and nights, warmer and more frequent hot days and nights	Increased yields in colder environments Decreased yields in warmer environments Increased insect outbreaks	Effects on water resources relying on snow melt Effects on some water supply	Reduced human mortality from decreased cold exposure	Reduced energy demand for heating and increased demand for cooling Declining air quality in cities Reduced disruption to transport due to snow, ice Effects on winter tourism
HEAT WAVES/ WARM SPELLS Frequency increases over most land areas	Reduced yields in warmer regions due to heat stress Wildfire danger increases	Increased water demand Water quality problems, e.g. algal blooms	Increased risk of heat-related mortality, especially for the elderly, chronically sick, very young and socially isolated	Reduction in quality of life for people in warm areas without appropriate housing Impacts on elderly, very young and poor
HEAVY PRECIPITATION EVENTS Frequency increases over most land areas	Damage to crops Soil erosion Inability to cultivate land due to waterlogging of soils	Adverse effects on quality of surface and groundwater Contamination of water supply Water stress may be relieved	Increased risk of deaths, injuries, infectious, respiratory and skin diseases	Disruption of settlements, commerce, transport and societies due to flooding Pressures on urban and rural infrastructures Loss of property
DROUGHT Affected areas increase	Land degradation Crop damage and failure Increased livestock deaths Increased risk of wildfire	More widespread water stress	Increased risk of malnutrition Increased risk of water- and food-borne diseases	Water shortages for settlements, industry and societies Reduced hydropower generation potentials
CYCLONES AND STORM SURGES Frequency increases	Damage to crops Windthrow (uprooting) of trees Damage to coral reefs	Power outages cause disruption of public water supply	Increased risk of deaths, injuries, water and food-borne diseases Posttraumatic stress disorders	Withdrawal of risk coverage in vulnerable areas by private insurers Potential for population migrations Loss of property
SEA LEVEL RISE Increased incidence of extreme high sea-level (excluding tsunamis)	Salinisation of irrigation water, estuaries and freshwater systems	Decreased freshwater availability due to salt-water intrusion	Increased risk of deaths and injuries by drowning in floods Migration-related health effects	Costs of coastal protection versus costs of land-use relocation Potential for movement of populations and infrastructure

Where are the most affected and who are the most vulnerable?

In all regions, both with low as well as high average incomes, some groups such as elderly people, the poor and young children, and systems or activities are more exposed than others. But there are sectors, systems and regions specially affected, namely the following:

On land:

At risk are areas that are particularly sensitive to the warming effect of climate change, including tundra, boreal forests of the North, and mountain regions. Others will be more affected by the reduced rainfall and changing precipitation patterns, in particular those with a Mediterranean-type environment and tropical rainforests.

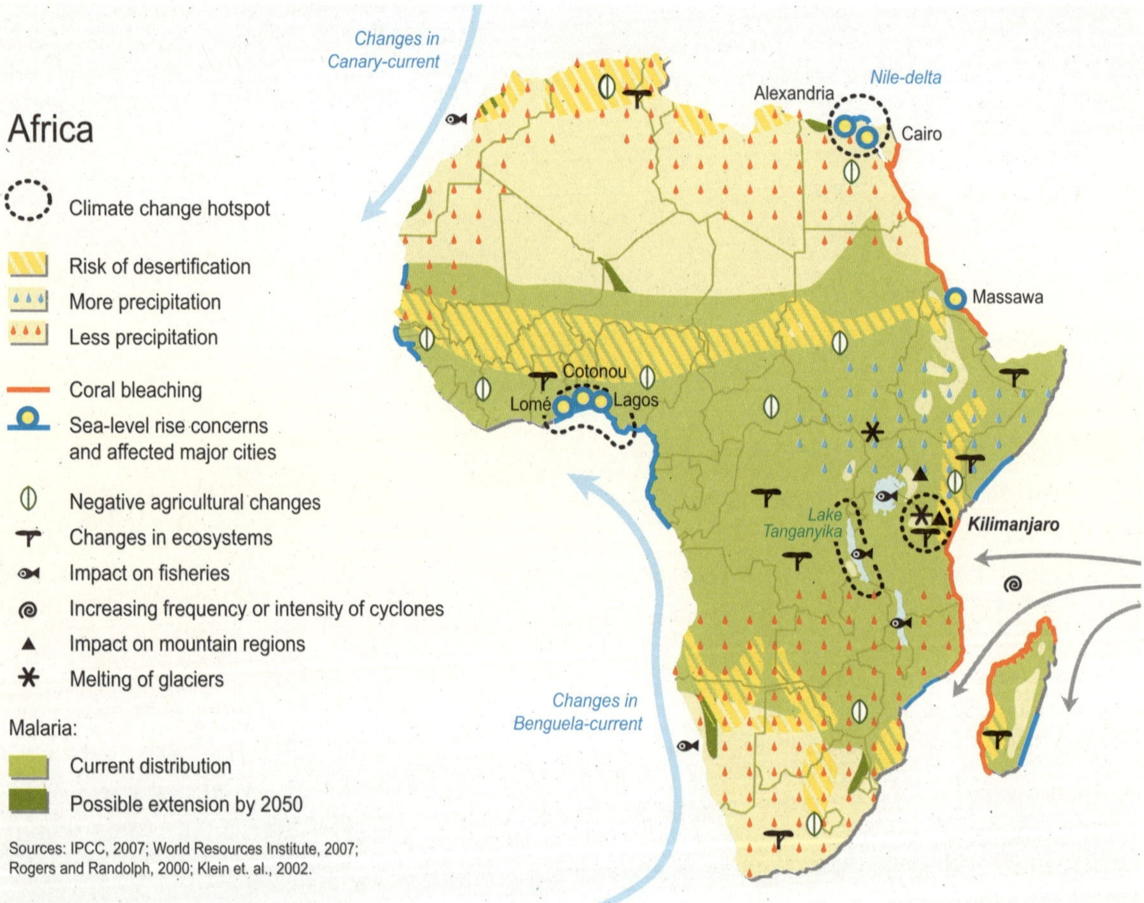

Sources: IPCC, 2007; World Resources Institute, 2007; Rogers and Randolph, 2000; Klein et. al., 2002.

On the coast:
Areas facing multiple stresses from climate change include mangroves and salt marshes.

In the ocean:
Coral reefs are ecosystems that are very vulnerable to thermal stress and find it hard to adapt. Increases in sea surface temperature of about 1–3°C are projected to result in more frequent coral bleaching and widespread mortality, unless there is thermal adaptation or acclimatization by corals. They face other climate-change related stresses such as increasing acidity and non climate-related ones (overfishing) in addition. Sea-ice biomes (communities) are also very sensitive to small changes in temperature, when ice turns to water for example.

Other risk sectors include low-lying coasts, water resources in some dry areas and those that depend on melting snow and ice, low-latitude agriculture, and human health in poorer countries. The increased frequency and intensity of extreme weather is "very likely" to worsen other impacts.

Regions expected to be especially hard-hit by climate change include:

The Arctic, because of the high rates of projected warming and its impact on people and the natural world. The thickness and extent of glaciers is expected to decline, and ice sheets and sea ice will also be affected. Invasive species may become a growing problem.

Africa, with the expected impacts on the continent and its low capacity for adaption. Yields from rain-fed agriculture, for example, could in some countries fall by half by 2020.

Small islands, where people and infrastructure are highly exposed to projected impacts, including sea level rise which is the main problem, as well as reduced rainfall in summer projected for these regions). This would reduce freshwater availability which may leave some unable to meet their demand. Increased rainfall in winter is "unlikely" to compensate this development due to lack of storage and high runoff during storms. In the Pacific, for example, a 10 per cent reduction in average rainfall (by 2050) would lead to a 20 per cent reduction of freshwater on

Population flooded in coastal areas in 2080

Million people per year (logarithmic scale)

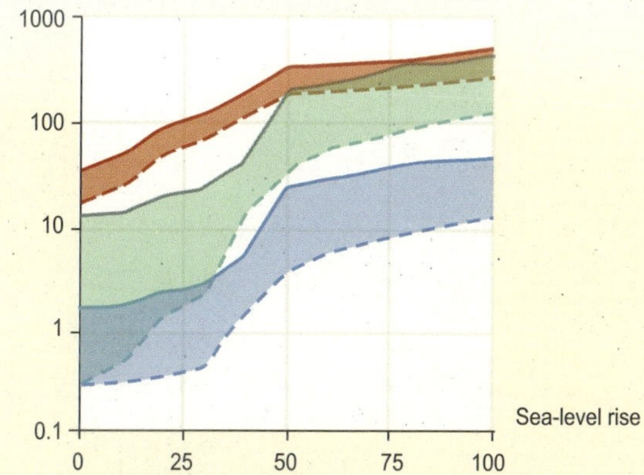

no additional efforts undetaken
more protection efforts than today
strong efforts to protect coastal populations against floods

Note: The upper margin of each band shows the amount of people affected in the A2 scenario according to which global population will reach 14 thousand million by 2080 with the lowest GDP of all IPCC scenarios. Therefore little capacity exists to adapt, and more people will be affected by floods. The lower end of each curve shows the impact for the A1/B1 scenario assuming the highest per captia income and world population at 8 thousand million, allowing for higher investments in the protection of the population.

Source: H. Ahlenius, GEO Ice and Snow, 2007, based on Nicholls, R.J. and Lowe, J.A., 2006.

Sources: IPCC, 2007; World Resources Institute, 2007; Klein et. al., 2002.

Tarawa Atoll, Kiribati. Furthermore, more alien species may take advantage of higher temperatures to settle on some islands, interfering with the natural ecosystems.

Asian and African mega-deltas, where large populations have to face high exposure to sea level rise, storm surges and river flooding. In east, south and south-east Asia illness and death from diarrhoeal disease caused by floods and droughts are expected to rise.

Elsewhere Australia and New Zealand are expected to face problems from reduced agricultural productivity and damage to species-rich areas, including the Great Barrier Reef.

Southern Europe may experience reduced availability of water, mountainous areas across the continent will face glacier retreat and reduced snow cover, leading to higher potential for water shortages, and health risks may increase from heat waves and wildfires.

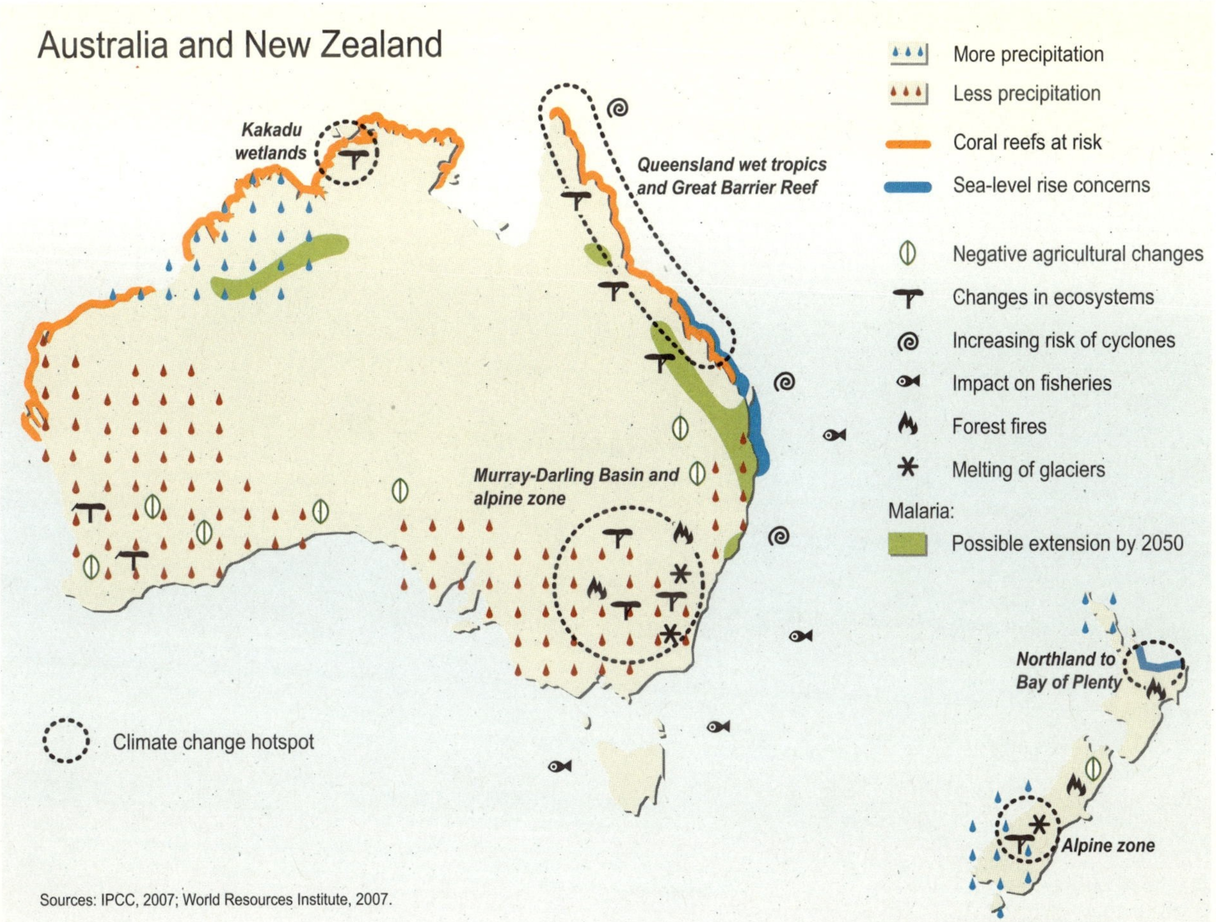

Europe

Climate change hotspot
Amsterdam
Rotterdam
Rhine delta
London
Hamburg
Alps
Venice
Istanbul
Mediterranean basin
More precipitation
Less precipitation
Sea-level rise concerns and affected major cities
Malaria:
Possible extension by 2050
Permafrost:
Present permafrost
Permafrost in 2050
Negative agricultural changes
Changes in ecosystems
Forest fires
Impact on fisheries
Impact on mountain regions
Melting of glaciers
Source: IPCC, 2007; Klein et. al., 2002.

Sources: IPCC, 2007; World Resources Institute, 2007; Rogers and Randolph, 2000; Klein et. al., 2002.

North America
More precipitation
Less precipitation
Sea-level rise concerns and affected major cities
Negative agricultural changes
Changes in ecosystems
Impact on fisheries
Increasing frequency or intensity of cyclones
Impact on mountain regions
Forest fires
Permafrost:
Present permafrost
Permafrost in 2050
Malaria:
Possible extension by 2050
Vancouver
New York
Los Angeles
New Orleans
Mississippi -delta
Climate change hotspot
Sources: IPCC, 2007; World Resources Institute, 2007; Rogers and Randolph; 2000; Klein et. al., 2002.

Latin America may have less water, as a consequence both of reduced precipitation and retreating glaciers, significant species loss, and by mid-century it may expect the gradual replacement of tropical forest by savanna in eastern Amazonia. Yields of food crops may diminish, exposing more people to the risk of hunger.

North America faces water scarcity, more heat waves, coastal threats and problems for some crops.

The risk of abrupt or irreversible changes

Climate warming could lead to some impacts that are abrupt or irreversible, depending upon the rate and magnitude of the change. Abrupt climate change on time scales of a decade or so is normally thought of as involving ocean circulation changes (like the meridional overturning circulation – see box). On longer time scales, ice sheet and ecosystem changes may also play a role.

If a large-scale abrupt climate change effect were to occur, its impact could be quite high. Partial loss of ice sheets on polar land and/or the thermal expansion of seawater over very long time scales could imply metres of sea level rise, with the greatest impacts on coasts, river deltas and islands, implying major changes in coastlines and inundation of low-lying areas. The number of people in the world who live within 100 km of the coast and no more than 100 m above sea level has been calculated at 600 millions to 1.2 billion – between 10 and 23 per cent of the world's population. Current models project that such changes would occur over very long time scales (millennia) if global temperature were to be sustained at 1.9–4.6°C over pre-industrial levels. Complete melting of the Greenland ice sheet would raise sea level by 7m and could be irreversible.

Meridional overturning circulation

The possibility which continues to fascinate the media, where potentially abrupt change could take place, is in the meridional overturning circulation of ocean water. Ocean currents have specific patterns that are determined by different water densities. Based on current model simulations, it is very likely that the *meridional overturning circulation* (MOC) of the Atlantic Ocean (the mixing of cold and warm water around the meridian) will slow down during this century (the Gulf Stream bringing warm water to Northern latitudes of Europe is part of this circulation system); nevertheless temperatures in the region are projected to increase. It is very unlikely that the MOC will undergo a large abrupt transition during the 21st century, and longer-term changes in the MOC cannot be confidently assessed. Impacts of large-scale and persistent changes in the MOC are likely to include changes in marine ecosystem productivity, fisheries, ocean CO_2 uptake, oceanic oxygen concentrations and terrestrial vegetation.[6]

6. About one-third of anthropogenic CO_2 emissions are thought to be entering the oceans, which form the largest active carbon "sink" – absorber – on Earth.

Regional impacts associated with global temperature change

Global mean annual temperature change relative to 1980 - 1999

Impacts will vary by extent of adaptation, rate of temperature change and socio-economic pathway

Adaptation and mitigation

Neither adaptation to climate change (reducing the potential impacts by changing the circumstances so that they strike less hard) nor mitigation (reducing the potential impacts by slowing down the process itself) alone can avoid all climate change impacts; however, they can complement each other and together significantly reduce the risks of climate change.

Adaptation is necessary both in the short and longer terms to address impacts resulting from the warming that would occur even if we make massive cuts in emissions. This is because the GHGs emitted already until today continue to have a warming effect on the climate, independent from how much the world continues to emit in the future. This was shown in chapter 2. We do however have some options to influence how big the future changes and their effects will be.

If nothing was done to slow climate change, it would, in the long term, be likely that natural, managed and even human systems would be unable to adapt and therefore fail to fulfill their purpose. The time at which such limits were reached would vary between sectors and regions. Early mitigation actions create the opportunity to develop alternatives for carbon-intensive infrastructure to reduce climate change and thereby also reduce the need (and the cost) for adaptation. Reliance on adaptation alone could eventually lead to climate change so severe that effective adaptation is not possible, or will be available only at very high social, environmental and economic costs.

Mitigation options

Mitigation of climate change seeks to reduce the rate and magnitude of climate change. Slowing down the processes of climate change would avoid or at least delay many of its impacts. If the amount of CO_2eq in the atmosphere is doubled over pre-industrial levels to around 550 ppm, the IPCC says global average temperatures will most probably rise at least 1.5°C or more. It cannot rule out the possibility that they could rise beyond 4.5°C.The next two to three decades will be crucial to achieve lower GHG stabilization levels, i.e. to stop the increase of GHGs concentrations in the atmosphere. If we fail to make the necessary efforts and investments and emissions reductions are delayed the risk of more severe climate change impacts increases. In order to stabilize the concentration of GHGs in the atmosphere, emissions would need to peak and decline thereafter. The lower the stabilization level we seek to achieve, the more quickly this peak and decline needs to occur.

In order to limit global warming to 2-2.4°C (2 degrees is the target set by the European Union and other countries, a temperature rise where it is believed that it would still be possible to adapt to its impacts with affordable efforts and costs) emissions would need to peak between 2000 and 2015. In reality, 2007 was a year of record increase in global CO_2 emissions.

Economic activities have a substantial potential for mitigation of GHG emissions over the coming decades. This could avoid the projected growth of global emissions, or even reduce them below their current levels. Some studies suggest mitigation opportunities with net gains. In other words, mitigation can create a positive financial result for the economy, for example through the development of new technologies or reduced energy costs.

It was estimated that already measures with net gains could reduce global CO_2eq emissions by 6 gigatonnes per year by 2030. Today, the global fossil fuel emissions are about 27 GtCO_2/yr.

Equally, there is high agreement among IPCC scientists and much evidence that all stabilization levels suggested in the table below can be achieved with either existing technology or new technologies that will be commercially available in the coming decades if appropriate framework conditions for their development are put in place.

Positive side-effects of early mitigation

There is a considerable potential for achieving a double benefit from measures to mitigate climate change. Reducing GHG emissions can result in large and rapid health benefits from reduced air pollution which may also offset a substantial part of the mitigation costs. Energy efficiency and the use of renewable energy offer synergies with sustainable development. In least-developed countries for example, changing the source

Stabilization scenarios

CO_2 concentration at stabilization[1]	CO_2-equivalent concentration at stabilization[2]	Peaking year for CO_2 emissions	Change in global CO_2 emissions in 2050 (% of 2000 emissions)	Global average temperature increase above pre-industrial at equilibrium	Global average sea level rise above pre-industrial at equilibrium from thermal expansion only[3]	Number of assessed scenarios
ppm	ppm	year	per cent	°C	metres	
350 - 400	445 - 490	2000 - 2015	-85 to -50	2.0 - 2.4	0.4 - 1.4	6
400 - 440	490 - 535	2000 - 2020	-60 to -30	2.4 - 2.8	0.5 - 1.7	18
440 - 485	535 - 590	2010 - 2030	-30 to +5	2.8 - 3.2	0.6 - 1.9	21
485 - 570	590 - 710	2020 - 2060	+10 to +60	3.2 - 4.0	0.6 - 2.4	118
570 - 660	710 - 855	2050 - 2080	+25 to +85	4.0 - 4.9	0.8 - 2.9	9
660 - 790	855 - 1130	2060 - 2090	+90 to +140	4.9 - 6.1	1.0 - 3.7	5

Note: the global average temperature at equilibrium is different from the expected global average temperature at the time of stabilization of GHG concentrations due to the inertia of the climate system. For the majority of scenarios assessed GHG concentrations are expected to stabilize between 2100 and 2150.

1 - Atmospheric CO_2 concentrations were 379ppm in 2005

2 - The best estimate of total CO2-eq concentration in 2005 for all long-lived GHGs is about 455ppm, while the corresponding value including the net effect of all anthropogenic forcing agents is 375ppm CO_2-eq

3 - Equilibrium sea level rise only reflects the contribution of ocean thermal expansion and does not reach equilibrium for many centuries. Long-term thermal expansion is projected to result in 0.2 to 0.6m per degree Celsius of global average warming above pre-industrial levels.

Mitigation techniques and strategies today and tomorrow

	Strategies available today	Strategies available tomorrow*
TRANSPORT	• hybrid and more fuel-efficient vehicles • cleaner diesel engines • biofuels • shifts from road to rail and public transport • cycling and walking • land use and transport planning	• second generation biofuels • higher-efficiency aircraft • advanced electric and hybrid vehicles with more powerful and reliable batteries
BUILDINGS	• more efficient appliances • efficient lighting • improved insulation • use of daylight • passive and active solar design • alternative refrigeration fluids • recovery and recycling of fluorinated gases	• integrated design of commercial buildings including technologies such as intelligent meters that provide feedback and control • solar photovoltaics integrated in buildings
INDUSTRY	• more efficient electrical equipment • heat and power recovery • material recycling and substitution • control of non-CO2 gas emissions • a wide array of process-specific manufacture	• advanced energy efficiency • CCS for cement, ammonia and iron manufacture • inert electrodes for aluminium
AGRICULTURE	• improved crop and pasture management to increase soil carbon storage • restoration of cultivated peaty soils and degraded land • improved rice-growing techniques and livestock and manure management to reduce methane emissions • improved nitrogen fertilizer	• Improved crop yields
FORESTRY	• afforestation and reforestation • reducing deforestation • improved management of forest resources • use of wood for bio-energy to replace fossil fuels	• tree species improvement to increase biomass productivity and carbon sequestration • improved remote sensing technology to analyse the carbon sequestration potential of vegetation and soils, and mapchanges in land-use
WASTE	• recovering methane from landfills • waste incineration with energy recovery • composting organic waste • controlled wastewater treatment • recycling and minimising waste	• biocovers and biofilters to optimize methane oxidization (such technologies improve methane combustion by minimizing carbon emissions).

*expected to be available commercially before 2030 according to the IPCC

of energy from wood to the sun can lower disease and death rates by cutting indoor air pollution, reducing the workload for women and children, who have to go out and collect the fuel wood, and decreasing the unsustainable use of fuel wood and therefore deforestation.

The role of policy and lifestyle

Policies that impose a real or implicit price on GHG emissions could create incentives for producers and consumers to invest significantly in products, technologies and processes generating few GHG emissions. An effective carbon-price signal could realize significant mitigation potential in all sectors. Modelling studies show that if global carbon prices were at 20 to 80 US$ per tonne of CO_2-eq by 2030 GHG concentrations in the atmosphere would stabilize at around 550ppm CO_2-eq by 2100. For the same stabilization level, other studies suggest these price ranges could be lowered to 5 to 65 US$ per tonne of CO_2-eq in 2030 as a result of changes in technologies that will take place by then.

Economic potential for reducing GHG emissions by sector in 2030

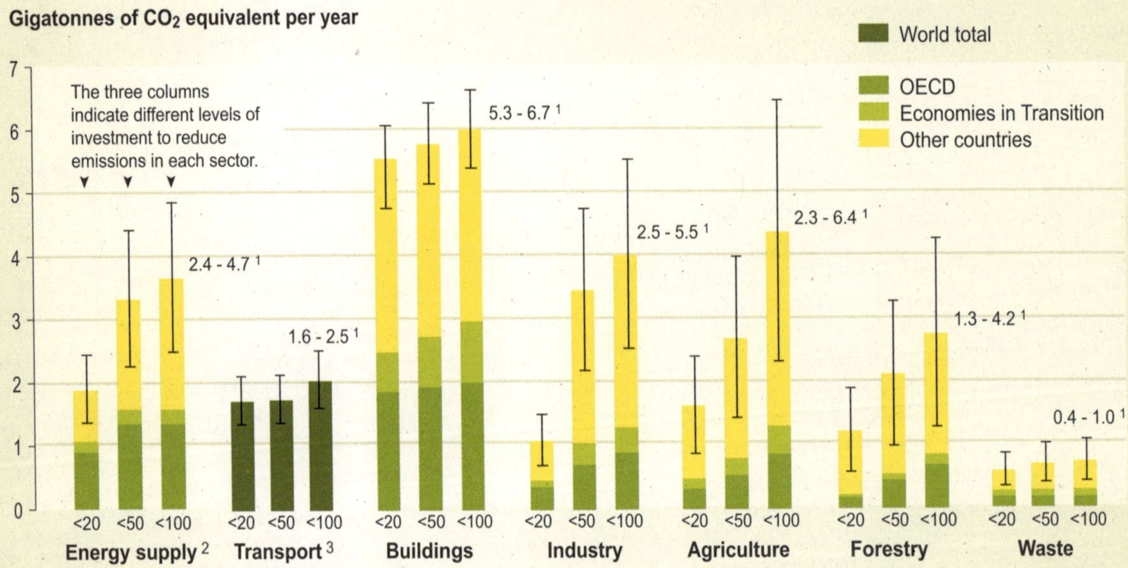

1. Total range of potential in each sector when investing up to US$100 per tonne of CO_2 eq
2. Electricity use is included in end use sectors, not in the energy sector.
3. As transport includes aviation, only global totals are shown.

Source: IPCC, 2007.

There is growing evidence that decisions about macroeconomic policy, for example agricultural policy, multilateral development bank lending, insurance practices, electricity market reform, energy security and forest conservation, which are often treated separately from climate policy, can reduce emissions significantly. Similarly, policies not directly linked to climate can affect both the capacity to adapt and vulnerability.

General findings about the performance of policies are:

- **Integrating climate policies in broader development policies** makes their implementation easier and helps to overcome barriers.
- **Regulations and standards** generally provide some certainty about emission levels. They may be preferable to other instruments when information or other barriers prevent producers and consumers from responding to price signals. However, they may not encourage innovations and more advanced technologies.
- **Taxes and charges** can set a price for carbon, but cannot guarantee a particular level of emissions. They can be an efficient way of internalizing the costs of GHG emissions.
- **Tradable permits** will establish a price for CO_2. The volume of allowed emissions determines their environmental effectiveness, while the allocation of permits has distributional consequences. Fluctuation in the price of CO_2 makes it difficult to estimate the total cost of complying with emission permits.
- **Financial incentives** (subsidies and tax credits) are frequently used by governments to stimulate the development and diffusion of new technologies. While economic costs are generally higher than for the options above, they are often critical to overcoming barriers.
- **Voluntary agreements** between industry and governments are politically attractive, raise awareness among stakeholders and have played a role in the evolution of many national policies. Most agreements have not achieved significant emissions reductions beyond business as usual. However, some recent agreements, in a few countries, have accelerated the application of best available technology and led to measurable emission reductions.
- **Information** (for example awareness campaigns) may improve environmental quality by promoting informed choices and possibly contributing to behavioural change, but the impact on emissions has not been measured yet.
- **Research, development and demonstration (RD&D)** can stimulate technological advances, reduce costs and help progress toward the stabilization of GHG concentrations.

Examples of ways to integrate climate change into development policies to encourage mitigation include:

- changes to taxes and subsidies so as to promote sustainable development;
- demand-side management programmes to reduce electricity consumption as well as cutting transmission and distribution losses;
- diversifying away from oil imports and reducing the economy's energy intensity;
- providing green incentives in the insurance of buildings and transport; and
- using the international finance system's country and sector strategies and project lending in a way that reduces emissions (by favouring low energy-intensity investments, for example).

Changes in lifestyle and behaviour patterns can contribute widely to climate change mitigation. Management practices can also have a positive role. Examples include changes in consumption patterns, education and training, changes in building occupant behaviour, transport demand management, and management tools in industry.

National / International scale

- **High-emitting facilities**
  - Cement factories
  - Iron and steel manufacturing
  - Chemical industries (ammonia, hydrogen, ethylene, ethanol, etc.)
  - Refineries
  - Gas processing plants
  - Power plants
  - Set performance standards ★
  - Control them ★
  - Switch to cleaner technologies
    - Carbon Capture and Storage ★
    - Nuclear ★★ ← Nuclear waste storage not resolved, remaining risks of nuclear accidents, uranium mining not clean
  - Subsidize possibilities of industrial reconversion for sectors emitting too much by nature ★

- **Transport**
  - Expand the national public transport network
  - Freight
    - Increase taxes for trucking freight
    - Develop rail and river freight alternatives ★
    - Advocate speed limits for international bunker freight shipping ★
  - Technological improvements
    - Vehicle design
      - Plane engines and wings
      - Hull shape and propellers of ships
      - Engines and exhaust systems of cars
    - Type of fuel
      - Biofuels ★★ ← Land use competition with food production, intensive agriculture
      - From heavy fuel oil to marine diesel oil for ships
  - Operational measures for ships and planes
    - Routing and timing for ships ★
    - Air traffic management (avoid waiting for landing) ★
  - Apply fossil fuel taxes to air traffic and shipping
  - Stop fossil fuel subsidies
  - Increase taxes on vehicle purchase and registration, and on use of roads and car parks ★ ← Equity problem: weighs more heavily on poor population

- **Agriculture and forestry**
  - Stop subsidies to intensive farming ★
  - Stop deforestation ★
  - Promote sustainable logging (labelling) ★
  - Control peat fires
  - Limit / control the use of fertilizers and pesticides
  - Produce for the local market ("food sovereignty" vs. global market)
  - Support sustainable farming practices

- **Set clear priorities in Research and Development allocation of funds**

  >> The text in boxes indicates priorities for research.

- **Energy**
  - Support renewable energy projects in all fields
    - Solar
    - Wind
    - Geothermal
    - Hydro-energy
    - Biomass
    - Ocean
  - Create incentives for energy efficiency improvement projects

- **Awareness-raising / educational campaigns**

- **GHG assessments of public sector and local authorities**

Political options

★ Controversial (orange star)

★ Needs international coordination

- Ratify and implement the Kyoto protocol > Commit to reduce national emissions
  - Enforce national reduction targets
  - Take part in international coordination programmes to reduce emissions (in transport, industry, etc.) ★
  - Offset unavoidable emissions ★
    - Planting trees (carbon sinks) ★ (controversial) — Carbon sequestration capacity of vegetation will attain maximum in a few decades
    - Finance projects in non-Annex I countries under the Clean Development Mechanism

Priority to local networks and diversification of energy sources
- Small loops for production, consumption, waste management ...
- Combine all local possibilities for clean energies

Local / City scale

- Construction
  - Establish sustainability criteria for buildings
  - Subsidize the construction of low-energy buildings
  - Subsidize insulations of existing buildings
  - Promote local and ecological building materials
  - Set public buildings as an example
- Waste management
  - "Less waste" policies
    - Support ecodesign projects (easy dismantling and recycling)
    - Support takeback campaigns
    - Organize sorting and recycling of waste
  - Energy recovery from waste
    - To heat buildings
    - To run industrial processes
- Urban planning
  - Limit urban sprawl
    - Subsidize collective housing in city centres
    - Subsidize the rehabilitation of unused or insalubrious buildings in city centres
    - Discourage real estate speculation in city centres
      - High housing tax for non-occupied building (office space in particular)
      - Use city and or state pre-emption rights to acquire land or buildings in town centres, for allocation to affordable collective housing
    - Make this goal a priority in official urban planning documents
  - Control and limit the use of cars in city centres
    - Develop pedestrian zones
    - Develop bicycle lanes and parks
    - Build car parks on city outskirts, close to public transport nodes
    - Widen pavements, making them easy for everyone to use (handicapped, strollers, etc.)
  - Decentralize and multiply service hubs (reducing the need for travel)
- Public transport
  - Expand the public transport network
  - Run a reliable, regular service (timetables, punctuality)
  - Make it affordable (subsidies, reduced prices)
  - Make it easy for everyone to use (handicapped, strollers, etc.)

Source: Emmanuelle Bournay, UNEP/GRID-Arendal inspired from the report "Mitigation of Climate Change", Working Group III, Fourth Assessment Report of the Intergovernmental Panel on Climate Change, 2007.

Same comfort level, different emissions

In this example, CO_2 emitted daily by two persons living in Munich are split up into the different activities which produce them, showing how easily less CO_2 could be emitted while enjoying the same level of comfort needs: 38 kg of CO_2 emitted by an average consumer versus 14 kg by a more aware one.

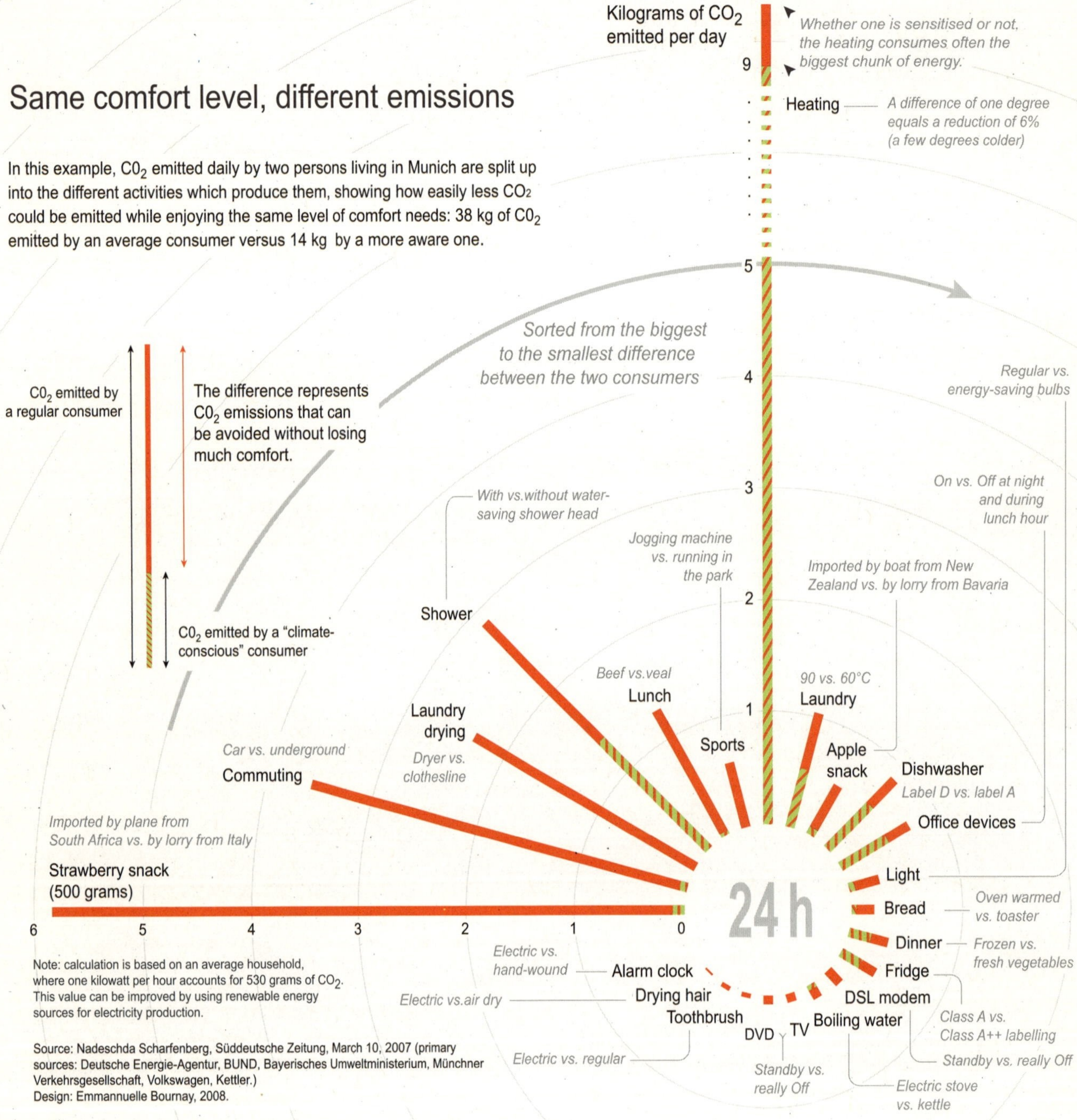

Note: calculation is based on an average household, where one kilowatt per hour accounts for 530 grams of CO_2. This value can be improved by using renewable energy sources for electricity production.

Source: Nadeschda Scharfenberg, Süddeutsche Zeitung, March 10, 2007 (primary sources: Deutsche Energie-Agentur, BUND, Bayerisches Umweltministerium, Münchner Verkehrsgesellschaft, Volkswagen, Kettler.)
Design: Emmannuelle Bournay, 2008.

Adaptation options

A wide range of options on how to adapt to a changing climate is available, but more adaptation than is happening now is needed to reduce vulnerability. Humans, other species, physical features and natural processes are all potentially vulnerable to the impacts of a changing climate, though to widely differing degrees. Vulnerability can be intensified by other stresses, for example poverty, hunger, trends in globalization, conflict, and diseases like HIV/Aids. The ability to adapt to the consequences of a warmer climate and thus to reduce its vulnerability to those depends closely on social and economic development and is unevenly distributed both within and between societies. But even societies with high adaptive capacity remain vulnerable to climate change, variability and extremes. For example, a heat wave in 2003 caused high levels of mortality in European cities (especially among the elderly), and Hurricane Katrina in 2005 caused large human and financial costs in the United States.

While little is known about the costs and benefits of adaptation at the global level, studies at the regional and project levels of impacts on specific sectors such as agriculture, energy demand for heating and cooling, water resources management and infrastructure are growing in number. These studies demonstrate that there are viable adaptation options available at low cost and/or with high benefit-cost ratios. Empirical research also suggests that higher benefit-cost ratios can be achieved by implementing some adaptation measures early, compared with retrofitting long-lived infrastructure at a later date.

Examples of adaptation strategies include:

- **water**: increased rainwater harvesting, water storage and conservation, water re-use, desalination, greater efficiency in water use and irrigation;
- **agriculture**: altering planting dates and crop varieties, relocating crops, better land management (for example erosion control and soil protection by planting trees);
- **infrastructure**: relocating people, building seawalls and storm surge barriers, reinforcing dunes, creating marshes and wetlands as buffers against sea level rise and floods;
- **human health**: action plans to cope with threats from extreme heat, emergency medical services, better climate-sensitive disease surveillance and control, safe water and improved sanitation;
- **tourism**: diversifying attractions and revenues, moving ski slopes to higher altitudes, artificial snow-making;
- **transport**: realigning and relocating routes, designing roads, railways and other transport equipment to cope with warming and drainage;
- **energy**: strengthening overhead transmission and distribution networks, putting some cabling underground, energy efficiency and renewable energy, reduced dependence on single energy sources.

The potential of international and regional cooperation

The United Nations Framework Convention on Climate Change (UNFCCC) and its Kyoto Protocol have made notable achievements in addressing the challenges climate change imposes on humanity. Responding globally to the climate change problem, stimulating an array of national policies, creating an international carbon market and establishing new institutional mechanisms may provide the foundation for future efforts to slow climate change. But to be more environmentally effective, future mitigation efforts would need to achieve deeper reductions covering activities with a higher share of global emissions.

Cooperation provides many options for achieving reductions of global GHG emissions at the international level. Successful agreements tend to be environmentally effective and cost-effective, to reflect the need for equity, and are institutionally feasible.

Efforts to address climate change can include diverse elements such as emissions targets; sectoral, local, sub-national and regional actions; research, development and demonstration (RD&D) programmes; adopting common policies; implementing development-oriented actions; or expanding financing instruments. These elements can be implemented in an integrated fashion, but comparing the efforts made by different countries quantitatively would be complex and resource-intensive.

Climate negotiations in the course of time

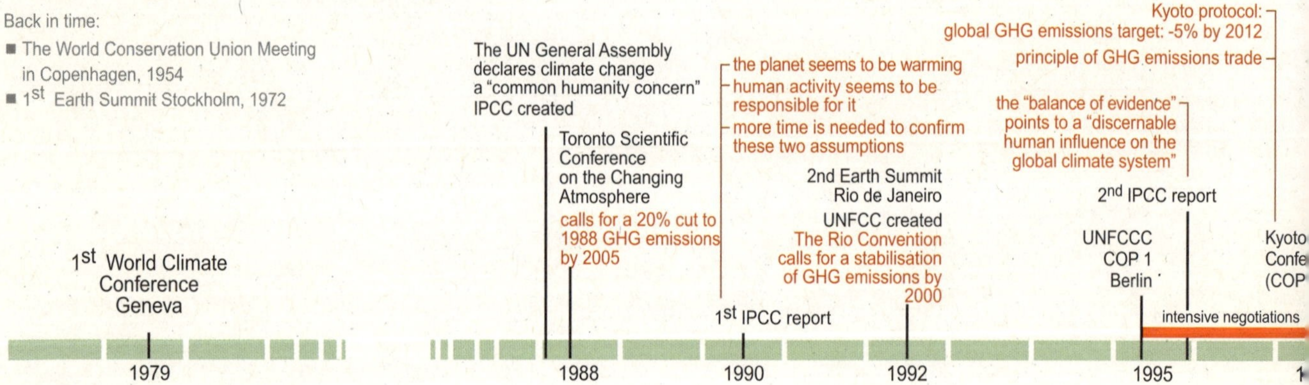

Sources: UNFCC, IPCC, Greenpeace. first published in: GRID-Arendal, Vital Climate Graphics, 2005.

The limits of adaptation and mitigation

Once GHG concentrations are stabilized, the rate at which the global average temperature increases is expected to slow within a few decades. But small temperature increases are expected for several centuries. Sea level rise from the expansion of warmer water would continue for many centuries at a gradually slowing rate, as the oceans will continue to absorb heat. The result would be an eventual sea level rise much larger than projected for the 21st century. Even if GHG and aerosol concentrations had been stabilized at the levels they had reached in 2000, thermal expansion alone would be expected to lead to further sea level rise of 0.3 to 0.8m.

This example shows how important it is that efforts to reduce the rate and magnitude of climate change and its impacts remember the inertia in both climate and socio-economic systems.

Adaptation will be ineffective in some cases such as natural ecosystems (e.g: the loss of Arctic sea ice and marine ecosystem viability), the disappearance of mountain glaciers that play vital roles in water storage and supply, or sea level rise of several metres. Adaptation will be more complicated or very costly in many cases for the climate change projected beyond the next few decades (for example in deltas and estuaries). The IPCC has found that the ability of many ecosystems to adapt naturally probably will be exceeded this century. In addition, multiple barriers and constraints to effective adaptation exist in human systems.

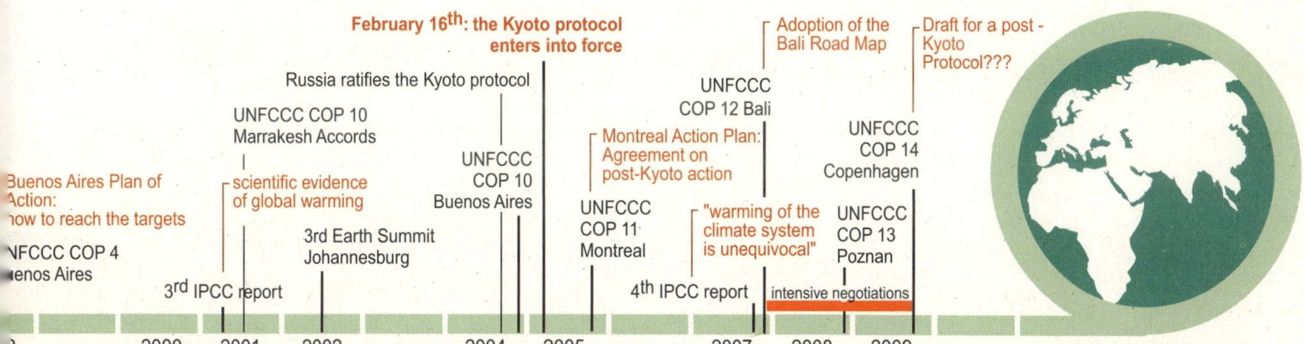

Costs of impacts, mitigation and long-term stabilization targets

Strategic options for climate change mitigation
Global cost curve for greenhouse gas abatement measures

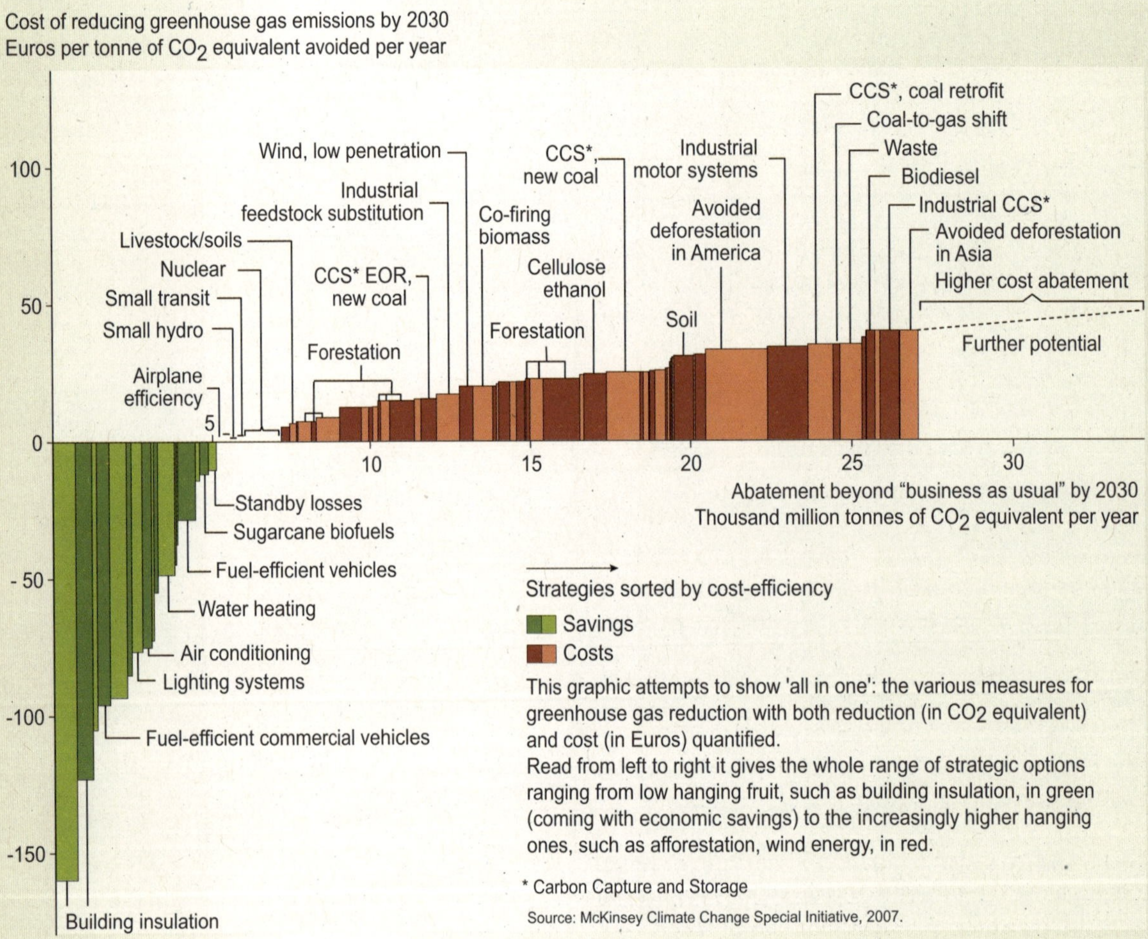

This graphic attempts to show 'all in one': the various measures for greenhouse gas reduction with both reduction (in CO_2 equivalent) and cost (in Euros) quantified.
Read from left to right it gives the whole range of strategic options ranging from low hanging fruit, such as building insulation, in green (coming with economic savings) to the increasingly higher hanging ones, such as afforestation, wind energy, in red.

* Carbon Capture and Storage

Source: McKinsey Climate Change Special Initiative, 2007.

The more severe the emission cuts to be made, the more it will usually cost. There is wide agreement that in 2050 global average costs for GHG mitigation towards stabilization between 710 and 445ppm of CO_2-eq are between a 1 per cent gain and a 5.5 per cent decrease in global GDP. This corresponds to slowing average annual global GDP growth by less than 0.12 percentage points, a rate that is much smaller than the annual fluctuation of GDP. GDP losses by 2030 are estimated at a cumulated 3 per cent since 2000 for the ambitious stabilization level of atmospheric GHGs between 445 and 535 ppm CO_2-eq. This corresponds to the estimated annual growth rate. If these predictions are correct, climate change mitigation would slow down global economic growth by one year by 2030.

For increases in global average temperature of less than 1–3°C above 1980–1999 levels, some impacts are projected to produce market benefits in some places and sectors while, at the same time, imposing costs in others. Global mean losses from impacts of climate change could be 1–5 per cent of GDP for 4°C of warming, but regional losses could be substantially higher. The social cost of carbon dioxide is the price of damages from climate change aggregated worldwide. This cost was estimated at 12 USD per tonne of CO_2 for 2005 and it is projected to increase over time. As these estimates do not include non-quantifiable impacts, they might underestimate real net costs.

Weighing the costs of mitigation against its benefits (that is, against the avoided impacts) does not as yet allow us to determine a stabilization level where benefits exceed costs.

Sustainable development, environmental protection and climate change

Climate change could impede nations' abilities to achieve sustainable development pathways. It is "very likely" that climate change can slow the pace of progress toward sustainable development either directly through increased exposure to adverse impacts or indirectly through erosion of the capacity to adapt. Over the next half-century, climate change could hamper the realization of the Millennium Development Goals. Climate change will interact at all scales with other trends in global environmental and natural resource problems, including water, soil and air pollution, health hazards, disaster risk, and deforestation. Their combined impacts may be compounded in future unless there are integrated mitigation and adaptation measures.

On the other side, mitigation of climate change can create synergies and avoid conflicts with other dimensions of sustainable development. For example, improving energy efficiency and developing renewable energies can improve energy security and reduce local pollution. Reducing deforestation will benefit biodiversity, and afforestation can restore degraded land, manage water runoff and benefit rural economies, if it does not compete with food production.

Similarly, sustainable development practices can enhance both the ability to adapt to and to mitigate climate change as well as reduce vulnerability to climate change.

Key vulnerabilities, impacts and risks – long-term perspectives

This section presents the essential climate change issues in brief, and compares recent findings with the state of knowledge of the Third Assessment Report which in 2001 identified five "reasons for concern" over the long term.

These "reasons" are judged to be stronger in the fourth report than previously. Many risks are identified with higher confidence. Some risks are projected to be larger or to occur at lower increases in temperature. Understanding about the relationship between impacts (the basis for "reasons for concern") and vulnerability (which includes the (in-)ability to adapt to impacts) has improved. This is because of more precise identification of the circumstances that make systems, sectors and regions especially vulnerable and growing evidence of the risks of very large impacts on multiple-century time scales.

Risks to unique and threatened systems. There is new and stronger evidence of observed impacts of climate change on unique and vulnerable systems (such as polar and high mountain communities and ecosystems), with increasing levels of adverse impacts as temperatures rise. An increasing risk of species extinction and coral reef damage is projected with higher confidence than in the TAR as warming proceeds.

Risks of extreme weather events. Responses to some recent extreme events reveal higher levels of vulnerability in both developing and developed countries than was assessed in the Third Assessment Report. There is now higher confidence in the projected increases in droughts, heat waves and floods.

Distribution of impacts and vulnerabilities. There are sharp differences across regions, and those in the weakest economic position are often the most vulnerable to climate change and are frequently the most susceptible to climate-related damages, especially when they face multiple stresses. There is increasing evidence of the greater vulnerability of specific groups such as the poor and elderly, not only in developing but also in developed countries. There is greater confidence in the projected regional patterns of climate change and in the projections of regional impacts, al-

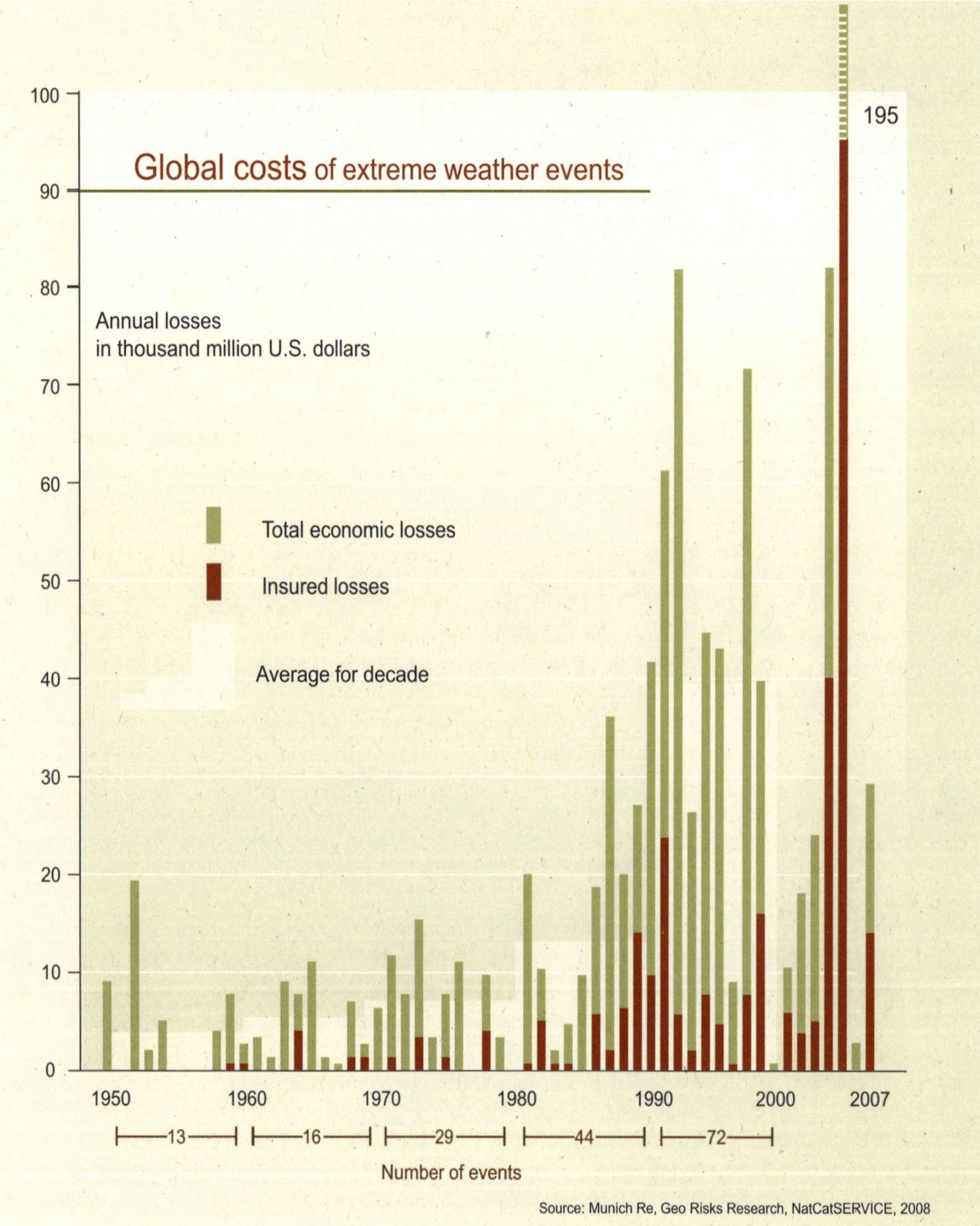

Global costs of extreme weather events
Annual losses
in thousand million U.S. dollars
Total economic losses
Insured losses
Average for decade
195
100
90
80
70
60
50
40
30
20
10
0
1950
1960
1970
1980
1990
2000
2007
13
16
29
44
72
Number of events
Source: Munich Re, Geo Risks Research, NatCatSERVICE, 2008

lowing better identification of particularly vulnerable systems, sectors and regions. And there is increased evidence that low-latitude and less-developed areas generally face greater risk, for example in dry areas and mega-deltas. New studies confirm that Africa is one of the most vulnerable continents because of the range of projected impacts, multiple stresses and low capacity to adapt. Substantial risks from sea level rise are projected, particularly for Asian mega-deltas and for small island communities.

Aggregate impacts. Compared with the third report, the initial market benefits of climate change are projected to peak at a lower temperature and therefore sooner than was assessed in the TAR. But economics are not the only way of quantifying the impacts of climate change: over the next century it is likely to adversely affect hundreds of millions of people through increased coastal flooding, reductions in water supplies, increased malnutrition and increased health impacts.

Risks of large-scale singularities (unique phenomena). During this century a large-scale abrupt change in the ocean circulation is very unlikely. Global warming over many centuries would lead to a sea level rise contribution from thermal expansion alone that is projected to be much larger than that observed over the 20th century, with the loss of coastal areas and associated impacts. There is better understanding than in the TAR that the risk of additional contributions to sea level rise from both the Greenland and possibly

Reasons for concern
about projected climate change impacts

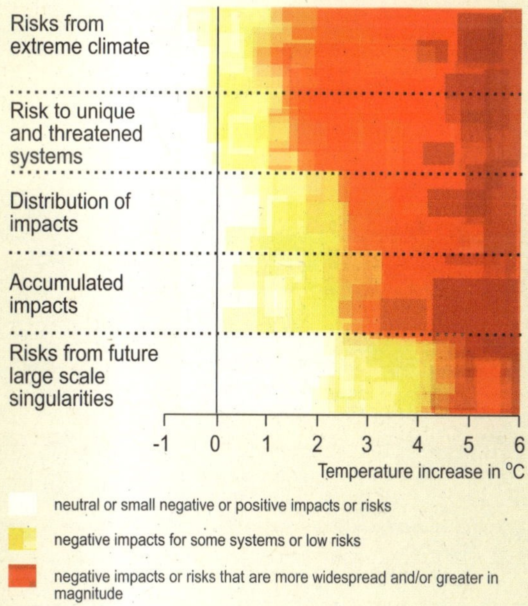

Source: IPCC, 2001.

Antarctic ice sheets may be larger than projected by ice sheet models and could occur on century time scales. This is because ice movements seen in recent observations, but not fully included in ice sheet models assessed in the Fourth Assessment Report, could increase the rate of ice loss.

Abbreviations and acronyms

AR4 The Fourth Assessment Report of the IPCC, published in 2007.

CO_2-eq Carbon dioxide-equivalent, used as measure for the emission (generally in $GtCO_2$-eq) or concentration (generally in ppm CO_2-eq) of GHGs.

Gigatonne One billion (one thousand million, or 10^9) tonnes.

GHGs Greenhouse gases. The six whose emissions are limited by the Kyoto Protocol are: Carbon dioxide (CO_2), methane (CH_4), nitrous oxide (N_20), sulphur hexafluoride (SF_6), HFCs (hydrofluorocarbons), PFCs (perfluorocarbons).

$GtCO_2$ Gigatonnes (metric) of carbon dioxide.

GWP Global Warming Potential: The combined effect of the length of time GHGs remain in the atmosphere and their effectiveness in trapping heat while they are there.

IPCC Intergovernmental Panel on Climate Change.

ppm Parts per million (as a measure of GHGs in the atmosphere).

TAR The IPCC's Third Assessment Report, published in 2001.

UNFCCC United Nations Framework Convention on Climate Change; see www.unfccc.int.